Angular Momentum Theory for Diatomic Molecules

BRIAN R. JUDD

Department of Physics
The Johns Hopkins University
Baltimore, Maryland

1975

ACADEMIC PRESS New York San Francisco London

A Subsidiary of Harcourt Brace Jovanovich, Publishers

ACADEMIC PRESS, INC.
111 Fifth Avenue, New York, New York 10003

United Kingdom Edition published by
ACADEMIC PRESS, INC. (LONDON) LTD.
24/28 Oval Road, London NW1

Library of Congress Cataloging in Publication Data

Judd, Brian R
Angular momentum theory for diatomic molecules.

Includes bibliographical references.
1. Molecular dynamics. 2. Angular momentum (Nuclear physics) 3. Chemistry, Physical and theoretical. I. Title.
QC175.16.M6J82 539'.6 74-1646
ISBN 0-12-391950-9

PRINTED IN THE UNITED STATES OF AMERICA

Angular Momentum Theory for Diatomic Molecules

Contents

6 Free Diatomic Molecules

7 The Hydrogen Molecule

8 External Fields

9 Perturbations

Preface

It has been recognized for a long time that angular momentum theory provides a convenient framework for describing the complex dynamical processes going on in a molecule. As early as 1927, Hund showed how molecular kinetics could be represented by the successive coupling of the angular momenta coming from the orbital motions and the spins of the constituent particles. Although methods for treating representations of this kind were rapidly developed, it is only comparatively recently that efforts have been made to put them in their most concise form. These techniques stem principally from Racah's work on tensor algebra in the 1940s. Their assimilation into atomic and molecular physics has taken place rather slowly, however. Even when used, Racah's methods are not always applied in the most effective or complete way. Parallel developments in other fields, such as nuclear physics, have had little impact on molecular structure, in spite of the existence of many common features.

One of the reasons for this state of affairs is that tensor algebra and molecular structure lie on opposite sides of the traditional demarcation line between physics and chemistry. This is reflected in several well-established texts on molecular structure, where apparently adequate descriptions of a wide variety of molecular phenomena are given without any mention of Racah. Although a number of review articles are now available in which this imbalance is corrected, there seems room for a general account of angular momentum theory, particularly Racah's innovations and their application to diatomic molecules. In fact, from a pedagogical point of view, molecules provide excellent opportunities for applications of tensor algebra. A course designed with this in mind could be profitably taken by students whose ultimate interests lie in other fields of physics or chemistry.

The successive coupling of various angular momenta provides a scheme for visualizing the kinetic structure of a molecule, but it would be complete only if spatial distributions could also be incorporated into the analysis. Some progress has been made in this direction for the hydrogen molecular ion. For more complicated diatomic molecules, however, a treatment based solely on angular momentum almost always concludes with the appearance of certain irreducible radial integrals. To evaluate them satisfactorily, modern computing techniques are required. This is a matter of theoretical chemistry, and is outside the scope of this book. The absence of precise theoretical values for the radial integrals is less of a shortcoming than might appear at first sight. Many experimental data can be correlated by regarding the integrals as adjustable parameters. If only a few parameters are required to do this, a rather satisfying understanding of the the molecular kinetics is obtained. At the same time, a natural break in the analysis occurs at the point where the evaluation of the parameters has to be faced. By limiting the analysis in this book to angular momentum techniques, it is possible to go quite deeply into this aspect of the theory of molecular structure and still hold the text to manageable proportions. Emphasis has been placed on the method of analysis, rather than on numerical results. Nothing approaching a comprehensive survey of diatomic molecules has been attempted.

Much of the material that follows has been presented to students and colleagues. Thanks go to Messrs. P. A. Lucas, B. I. Dunlap, and E. E. Vogel for checking various calculations. Comments from many physicists and chemists on points of detail are gratefully acknowledged. It is a pleasure to thank Dr. J. M. Brown and Professor A. Carrington for a preview of the early chapters of their projected book on molecular structure.

Angular Momentum Theory for Diatomic Molecules

1 Tensor Algebra

1.1 INTRODUCTION

In the years since the introduction of tensor operators, the theory of angular momentum and its application to quantum mechanics have expanded enormously. Some idea of the extent of the developments can be gained by comparing an early text, such as that of Edmonds [1], with the later book of Jucys and Bandzaitis [2], where elaborate graphical techniques are used to cope with the complex coupling schemes. For most applications to molecules, however, many of the later developments are unnecessarily sophisticated. The aim of the present chapter is to summarize the relevant parts of the theory and to prepare the ground for subsequent applications.

The close connection between an angular momentum vector and a rotation operator provides a basis for interpreting the theory in terms of the rotation group in three dimensions, $R(3)$. Although it will be assumed that the reader is somewhat familiar with group theoretical concepts—if only those of finite groups—we will only exploit the underlying group structure of a problem if a significant improvement in comprehension or insight is achieved. This parallels the approach of the two books mentioned above, as well as that of Brink and Satchler [3]. That it works at all stems from the special nature of $R(3)$, whose representations, unlike those of most groups, are susceptible of a rather complete algebraic description.

1.2 COMMUTATION RELATIONS

In quantum mechanics, the components of a position vector $\mathbf{r}$ of a particle and those of its momentum $\mathbf{p}$ satisfy certain basic commutation relations. For notational convenience, we extract $\hbar$ from $\mathbf{p}$, so that the momentum of the particle becomes $\hbar\mathbf{p}$. The commutation relations now run

$$[x, p_x] = i, \qquad [x, p_y] = [x, p_z] = 0, \tag{1.1}$$

together with all cyclic permutations. The notation $[a, b] = ab - ba$ is used here. The angular momentum $\boldsymbol{l}$ of the particle, measured in units of $\hbar$, is defined as

$$\boldsymbol{l} = \mathbf{r} \times \mathbf{p}.$$

Owing to the basic commutation relations (1.1), the components of $\boldsymbol{l}$ satisfy the equations

$$[l_x, l_y] = il_z, \qquad [l_y, l_z] = il_x, \qquad [l_z, l_x] = il_y.$$

It is postulated that the components of any angular momentum vector $\mathbf{J}$ (which may have contributions coming from spins) satisfy the analogous equations

$$[J_x, J_y] = iJ_z, \qquad [J_y, J_z] = iJ_x, \qquad [J_z, J_x] = iJ_y. \tag{1.2}$$

It is often convenient to use the linear combinations $J_\pm = J_x \pm iJ_y$, in which case Eqs. (1.2) become

$$[J_z, J_\pm] = \pm J_\pm, \qquad [J_+, J_-] = 2J_z.$$

Associated with $\mathbf{J}$ are the kets $|J, M\rangle$ for which we may write

$$J_z\,|J, M\rangle = M\,|J, M\rangle,$$

$$\mathbf{J}^2\,|J, M\rangle = J(J+1)\,|J, M\rangle,$$

$$J_\pm\,|J, M\rangle = [J(J+1) - M(M \pm 1)]^{1/2}\,|J, M \pm 1\rangle. \tag{1.3}$$

The symbols J and M are quantum numbers, and are restricted to either integral or half-integral values. The kets are orthonormal; that is,

$$\langle J, M \mid J', M' \rangle = \delta(M, M')\delta(J, J').$$

It might be useful to introduce here an explicit representation for **J** and the kets. To this end, we bring into play the spherical polar coordinates $(r\theta\phi)$ by means of the equations

$$x = r \sin\theta \cos\phi, \qquad y = r\sin\theta\sin\phi, \qquad z = \cos\theta,$$

and then we define the spherical harmonics Y_{lm} as

$$Y_{lm}(\theta, \phi) = (-1)^m \left\{\frac{(2l+1)(l-m)!}{4\pi(l+m)!}\right\}^{1/2} P_l^m(\cos\theta)e^{im\phi}, \quad (1.4)$$

where

$$P_l^m(\mu) = \frac{(1-\mu^2)^{m/2}}{2^l l!}\frac{d^{l+m}}{d\mu^{l+m}}(\mu^2-1)^l \qquad (|\mu| \leq 1). \quad (1.5)$$

In these equations, m runs over all integers, both positive and negative, in the interval $-l$ to $+l$. The angle θ is restricted to the range $0 \leq \theta \leq \pi$. If we now use the familiar representations

$$p_x = \frac{1}{i}\frac{\partial}{\partial x},$$

etc., and then pass to polar coordinates, we quickly find that

$$l_z = \frac{1}{i}\frac{\partial}{\partial\phi}, \qquad l_\pm = e^{\pm i\phi}\left(\pm\frac{\partial}{\partial\theta} + i\cot\theta\frac{\partial}{\partial\phi}\right). \quad (1.6)$$

It is now a straightforward matter to show that Eqs. (1.3) are valid if we make the correspondences

$$\mathbf{J} \to \mathbf{l}, \qquad J \to l, \qquad M \to m, \qquad |J, M\rangle \to Y_{lm}.$$

The orthonormality

$$\int Y_{lm}^* Y_{l'm'}\, d\tau = \delta(m, m')\delta(l, l')$$

is assured if we take $d\tau = \sin\theta\, d\theta\, d\phi$ and limit θ and ϕ to the regions $0 \leq \theta \leq \pi$, $0 \leq \phi \leq 2\pi$. In this way the surface of the unit sphere is covered just once in the integration.

1.3 ROTATIONS

One of the most remarkable features of angular momentum vectors is their usefulness in constructing rotation operators. Consider, for example,

$$\exp(-i\alpha l_z)\ (x \pm iy)\ \exp(i\alpha l_z).$$

To evaluate this quantity [4], we take as our starting point

$$l_z(x \pm iy) = (x \pm iy)(l_z \pm 1).$$

Thus

$$l_z^2(x \pm iy) = l_z(x \pm iy)(l_z \pm 1) = (x \pm iy)(l_z \pm 1)^2,$$

and, in general,

$$l_z^n(x \pm iy) = (x \pm iy)(l_z \pm 1)^n.$$

On expanding the exponential, we at once see that

$$\exp(-i\alpha l_z)\ (x \pm iy) = (x \pm iy)\ \exp(-i\alpha(l_z \pm 1)),$$

and so

$$\exp(-i\alpha l_z)\ (x \pm iy)\ \exp(i\alpha l_z) = (x \pm iy)\ \exp(\mp i\alpha). \tag{1.7}$$

Since l_z commutes with z, we can immediately write down

$$\exp(-i\alpha l_z)\ z\ \exp(i\alpha l_z) = z. \tag{1.8}$$

In many cases, the algebra of the transformation may be sufficient for our purposes. However, if we wish to put a geometrical interpretation on the equations, two options are open to us. We can imagine either that the operator $\exp(-i\alpha l_z)$ rotates the vector $\mathbf{r}$ through an angle $-\alpha$ about the z axis, or else that it rotates the coordinate frame by an angle $+\alpha$ about the axis. To make sure that all spins and momenta are similarly transformed, we have merely to use $\exp(-i\alpha J_z)$, where $\mathbf{J}$ is the total angular momentum of the system.

Suppose that two rectangular coordinate frames F and F' share the same origin but are arbitrarily oriented, one with respect to the other. To bring F into coincidence with F', a single rotation will, of course, suffice; but if the allowed axes of rotation are specified in advance, then three successive rotations are, in general, necessary. If we rotate F first by γ about its z axis, then by β about the y axis of F (i.e., about the original y axis, not the new one), and finally by α about the z axis of F (i.e., about the same axis as the first rotation), the rotation operator is

$$D(\omega) = \exp(-i\alpha J_z)\ \exp(-i\beta J_y)\ \exp(-i\gamma J_z), \tag{1.9}$$

where ω is an abbreviation for the three Euler angles α, β, and γ. Let (xyz)

be the coordinates of a point in F and $(\xi\eta\zeta)$ its coordinates in F'. Then

$$\xi = D(\omega)xD(\omega)^{-1},$$

with similar equations for η and ζ. For future reference, the results are written out in full:

$$\begin{aligned}\xi &= x(\cos\alpha\cos\beta\cos\gamma - \sin\alpha\sin\gamma)\\ &\quad + y(\sin\alpha\cos\beta\cos\gamma + \cos\alpha\sin\gamma) - z\sin\beta\cos\gamma,\\ \eta &= x(-\cos\alpha\cos\beta\sin\gamma - \sin\alpha\cos\gamma)\\ &\quad + y(-\sin\alpha\cos\beta\sin\gamma + \cos\alpha\cos\gamma) + z\sin\beta\sin\gamma,\\ \zeta &= x\cos\alpha\sin\beta + y\sin\alpha\sin\beta + z\cos\beta. \end{aligned} \tag{1.10}$$

The inverses can be written down by making the replacements

$$x \to \xi, \quad y \to \eta, \quad z \to \zeta, \quad \alpha \to -\gamma, \quad \beta \to -\beta, \quad \gamma \to -\alpha.$$

All these equations can be obtained by repeated use of Eqs. (1.7)–(1.8) and cyclic permutations of them. In carrying through the calculations, we must avoid using equations such as

$$e^a e^b = e^{a+b} = e^{b+a} = e^b e^a,$$

which only hold if a and b commute. (The correct form of equations of this type has been discussed by Messiah [5].)

There exists an alternative way of making the transformation above. First, F is rotated by α about its z axis, then this second frame (say F'') is rotated by β about the new y axis. Finally, this third frame is rotated by γ about its z axis. A detailed analysis reveals that the frame so obtained coincides with F' (see Problem 1.1).

1.4 ROTATION MATRICES

Corresponding to the coordinate transformations (1.10), any operator T becomes T', where

$$T' = D(\omega)TD(\omega)^{-1}.$$

The effect of $D(\omega)$ on a ket $|J, M'\rangle$ can only be to produce a linear combination of kets with the same J, since $D(\omega)$ is a function of the components of $\mathbf{J}$, for which Eqs. (1.3) are valid. Let us therefore write

$$D(\omega)\,|JM'\rangle = \sum_{M''} \mathfrak{D}_{M''M'}{}^{J}(\omega)\,|JM''\rangle$$

and determine the numerical coefficients $\mathfrak{D}_{M''M'}{}^{J}(\omega)$. (For simplicity, the commas have been omitted from the kets.) Setting the bra $\langle JM|$ to the left of both sides of this equation, and writing $M' = N$, we get

$$\begin{aligned}\langle JM|D(\omega)|JN\rangle &= \mathfrak{D}_{MN}{}^J(\omega)\\ &= \langle JM|\exp(-i\alpha J_z)\exp(-i\beta J_y)\exp(-i\gamma J_z)|JN\rangle\\ &= \exp(-i(\alpha M + \gamma N))\,d_{MN}{}^J(\beta),\end{aligned}$$

where

$$d_{MN}{}^J(\beta) = \langle JM|\exp(-i\beta J_y)|JN\rangle.$$

The derivation of the explicit form of $d_{MN}{}^J(\beta)$ is outlined in Problem (1.2); the final answer is

$$d_{MN}{}^J(\beta) = \sum_t (-1)^t \frac{[(J+M)!(J-M)!(J+N)!(J-N)!]^{1/2}}{(J+M-t)!(J-N-t)!t!(t+N-M)!} \times (\cos\beta/2)^{2J+M-N-2t}(\sin\beta/2)^{2t+N-M}. \tag{1.11}$$

This expression for $d_{MN}{}^J(\beta)$ enables many symmetry properties of the rotation matrices to be rapidly written down. For example,

$$d_{MN}{}^J(\beta) = d_{-N-M}{}^J(\beta), \tag{1.12}$$

$$d_{NM}{}^J(\beta) = d_{MN}{}^J(-\beta) = (-1)^{M-N}d_{MN}{}^J(\beta), \tag{1.13}$$

from which we can deduce

$$\mathfrak{D}_{MN}{}^J(\omega)^* = (-1)^{M-N}\mathfrak{D}_{-M-N}{}^J(\omega). \tag{1.14}$$

An expression such as $d_{MN}{}^J(\pi-\beta)$ is slightly more difficult to handle. However, we have only to write $t = J - t' - N$, and the sum over t' can be expressed in terms of $d_{-MN}{}^J(\beta)$. The result is

$$d_{MN}{}^J(\pi-\beta) = (-1)^{J-N}d_{-MN}{}^J(\beta) = (-1)^{J+M}d_{M-N}{}^J(\beta). \tag{1.15}$$

If we recall that the adjoint of an operator product AB is given by $(AB)^\dagger = B^\dagger A^\dagger$, it is at once seen that $D(\omega)^\dagger = D(\omega)^{-1}$, since the Cartesian components of $\mathbf{J}$ are Hermitian [6]. It follows that $D(\omega)$ is a unitary operator, and that the matrices $\mathfrak{D}_{MN}{}^J(\omega)$, where M and N label the rows and columns respectively, are unitary matrices. Thus

$$\begin{aligned}&\sum_{M'} \mathfrak{D}_{M'N}{}^J(\omega)^*\mathfrak{D}_{M'M}{}^J(\omega) = \delta(M,N),\\ &\sum_{M'} \mathfrak{D}_{MM'}{}^J(\omega)\mathfrak{D}_{NM'}{}^J(\omega)^* = \delta(M,N).\end{aligned} \tag{1.16}$$

If a ket can be represented by a wavefunction $\psi(\mathbf{r})$, then the effect of $D(\omega)$ is to rotate the contours of $\psi(\mathbf{r})$. However, the sense of the rotation is opposite to that for **r** itself. To see this, take $\psi(\mathbf{r})$ in the hypothetical form of the Dirac delta function $\delta(\mathbf{r} - \mathbf{R})$. Suppose that **r** becomes $\mathbf{r} + \mathbf{a}$ under the action of $D(\omega)$. Then the wavefunction becomes $\delta(\mathbf{r} + \mathbf{a} - \mathbf{R})$, and the singularity moves from **R** to $\mathbf{R} - \mathbf{a}$.

1.5 SPHERICAL TENSORS

Operators that transform under rotations in the same manner as the kets $|JM\rangle$ are of great importance. If the $2k + 1$ operators $T_q^{(k)}$, where $-k \leq q \leq k$, satisfy

$$D(\omega)T_q^{(k)}D(\omega)^{-1} = \sum_{q'} \mathfrak{D}_{q'q}{}^k(\omega)T_{q'}{}^{(k)}, \tag{1.17}$$

then they are said to form the components of a spherical tensor $\mathbf{T}^{(k)}$ of rank k. It is clear that the commutation relations that the $T_q^{(k)}$ obey with respect to **J** are of crucial importance in determining the form of the right-hand side of Eq. (1.17). In fact, we can follow Racah [7] and use these commutation relations as an equivalent way of defining a tensor operator. They run

$$[J_z, T_q^{(k)}] = qT_q^{(k)},$$
$$[J_\pm, T_q^{(k)}] = [k(k+1) - q(q \pm 1)]^{1/2}T_{q\pm1}{}^{(k)}. \tag{1.18}$$

The similarity between these conditions and Eqs. (1.3) illustrates the correspondence between tensor operators and kets.

The vector **J** is itself a tensor operator. To satisfy Eqs. (1.18), we must define the components as follows:

$$J_1^{(1)} = -\sqrt{\tfrac{1}{2}}\,J_+, \qquad J_0^{(1)} = J_z, \qquad J_{-1}^{(1)} = \sqrt{\tfrac{1}{2}}\,J_-.$$

No adjustment needs to be made for the spherical harmonics Y_{lm}; they are tensors for which $k = l$ and $q = m$. It is often more convenient to use the tensors $\mathbf{C}^{(k)}$, for which

$$C_q^{(k)}(\theta\phi) = \sqrt{\frac{4\pi}{2k+1}}\, Y_{kq}(\theta, \phi).$$

It is found, for example, that

$$C_{\pm1}^{(1)} = \mp\sqrt{\frac{1}{2}}\,\frac{x \pm iy}{r}, \qquad C_0^{(1)} = \frac{z}{r}. \tag{1.19}$$

Here and elsewhere, the radical sign applies only to the fraction immediately following it.

1.6 DOUBLE TENSORS

In the previous sections we have specified $D(\omega)$ in terms of the components of the total angular momentum $\mathbf{J}$, and it is through the commutation relations with this operator that tensor operators have been defined. It is often useful, however, to consider the properties of operators with respect to other angular momenta before examining their features overall. Commuting angular momenta turn out to be particularly valuable for separating and characterizing the properties of operators in the different spaces spanned by the angular momenta. The orbital and spin spaces, to which correspond the total orbital and total spin angular momentum respectively, constitute perhaps the most familiar example. If we have two commuting angular momenta $\mathbf{J}$ and $\mathbf{J}'$, then a double tensor $\mathbf{T}^{(kk')}$ can be defined by specifying that the $(2k + 1)$ components $T_{qq'}{}^{(kk')}$ for which $-k \leq q \leq k$ behave as a tensor with respect to $\mathbf{J}$ for any choice of q'; and, reciprocally, the $(2k' + 1)$ components $T_{qq'}{}^{(kk')}$ for which $-k' \leq q' \leq k'$ behave as a tensor with respect to $\mathbf{J}'$ for any choice of q. The double tensor $\mathbf{T}^{(kk')}$ possesses $(2k + 1)(2k' + 1)$ components in all.

An important example of $\mathbf{J}$ and $\mathbf{J}'$ can be constructed from the coordinates (xyz) and $(\xi\eta\zeta)$. Since the nine quantities $(xyz,\ \xi\eta\zeta,\ \alpha\beta\gamma)$ are connected by just three equations—namely, Eqs. (1.10)—any six of them can be taken as independent variables. If $(xyz,\ \xi\eta\zeta)$ are chosen for this role, we can define two angular momentum vectors $\boldsymbol{l}$ and $\boldsymbol{\lambda}$ by means of the equations

$$l_x = \frac{1}{i}\left(y\frac{\partial}{\partial z} - z\frac{\partial}{\partial y}\right), \tag{1.20}$$

$$\lambda_\xi = \frac{1}{i}\left(\eta\frac{\partial}{\partial \zeta} - \zeta\frac{\partial}{\partial \eta}\right), \tag{1.21}$$

together with their cyclic permutations over the respective triads (xyz) and $(\xi\eta\zeta)$. Having taken $(xyz,\ \xi\eta\zeta)$ as independent variables, we know that

$$[\boldsymbol{l}, \boldsymbol{\lambda}] = 0,$$

so we can make the identifications $\boldsymbol{l} = \mathbf{J}$, $\boldsymbol{\lambda} = \mathbf{J}'$.

A physical interpretation can be given to $\boldsymbol{l}$ and $\boldsymbol{\lambda}$. The vector $\boldsymbol{l}$ is evidently the orbital angular momentum of a particle measured with respect to a frame F; but since the partial derivatives $\partial/\partial x$, $\partial/\partial y$, and $\partial/\partial z$ that occur in $\boldsymbol{l}$ imply differentiation in which ξ, η, and ζ are held constant, the particle is stationary in F'. In other words, the frame F' turns relative to F in such a way that the motion of the particle is precisely followed. We may evidently express $\boldsymbol{l}$ in terms of the Euler angles through which the relative position of F and F' is defined. When this is done, we get

$$l_x \pm il_y = ie^{\pm i\alpha}\left(\cot\beta\,\frac{\partial}{\partial\alpha} \mp i\,\frac{\partial}{\partial\beta} - \frac{1}{\sin\beta}\,\frac{\partial}{\partial\gamma}\right),$$

$$l_z = -i\,\frac{\partial}{\partial\alpha}. \tag{1.22}$$

In a similar way, $\boldsymbol{\lambda}$ represents the orbital angular momentum of a particle which is stationary in F. For this case,

$$\lambda_\xi \pm i\lambda_\eta = ie^{\mp i\gamma}\left(\cot\beta\,\frac{\partial}{\partial\gamma} \pm i\,\frac{\partial}{\partial\beta} - \frac{1}{\sin\beta}\,\frac{\partial}{\partial\alpha}\right),$$

$$\lambda_\zeta = i\,\frac{\partial}{\partial\gamma}. \tag{1.23}$$

The partial derivatives $\partial/\partial\alpha$, $\partial/\partial\beta$, and $\partial/\partial\gamma$ in Eqs. (1.22) are not the same as those in Eqs. (1.23): the former correspond to a scheme in which $(\xi\eta\zeta)$ are held constant, the latter to one in which (xyz) are held constant. If we wish to use $\boldsymbol{l}$ and $\boldsymbol{\lambda}$ to investigate the tensorial properties of functions solely of the Euler angles, this distinction is irrelevant. Without further ado, we can take the right-hand sides of Eqs. (1.22) and (1.23) for two commuting operators $\boldsymbol{l}$ and $\boldsymbol{\lambda}$.

Consider, for example, $\mathfrak{D}_{MN}{}^J(\omega)^*$. We see at once that

$$[l_z, \mathfrak{D}_{MN}{}^J(\omega)^*] = M\mathfrak{D}_{MN}{}^J(\omega)^*, \tag{1.24}$$

since the dependence of the $\mathfrak{D}^*$ function on α is simply $e^{iM\alpha}$. The commutation with respect to $l_\pm$ is more difficult to work out, though quite straightforward if $\cot\beta$ and $\csc\beta$ are converted to functions of half angles through the expressions

$$\cot\beta = \frac{\cos^2\beta/2 - \sin^2\beta/2}{2\sin\beta/2\cos\beta/2}, \qquad \csc\beta = \frac{\cos^2\beta/2 + \sin^2\beta/2}{2\sin\beta/2\cos\beta/2}$$

before combination with the expression for $d_{MN}{}^J(\beta)$ given in Eq. (1.11). The result is

$$[l_x \pm il_y, \mathfrak{D}_{MN}{}^J(\omega)^*] = [J(J+1) - M(M\pm 1)]^{1/2}\mathfrak{D}_{M\pm1,N}{}^J(\omega)^*. \tag{1.25}$$

So, for a given N (and ω), the functions $\mathfrak{D}_{MN}{}^J(\omega)^*$ behave as tensor components $T_M{}^{(J)}$ with respect to $\boldsymbol{l}$.

When we turn to $\boldsymbol{\lambda}$, a slight complication arises. The commutation relations turn out to be

$$[\lambda_\zeta, \mathfrak{D}_{MN}{}^J(\omega)^*] = -N\mathfrak{D}_{MN}{}^J(\omega)^*,$$

$$[\lambda_\xi \pm i\lambda_\eta, \mathfrak{D}_{MN}{}^J(\omega)^*] = -[J(J+1) - N(N \mp 1)]^{1/2}\mathfrak{D}_{M,N\mp1}{}^J(\omega)^*. \tag{1.26}$$

To get a minus sign in the first equation, we need to assign a tensor component q of $-N$ to $\mathfrak{D}_{MN}{}^J(\omega)^*$. The minus sign in the second equation requires an alternation of sign with N for the tensor components. This is most conveniently achieved by means of a phase factor $(-1)^{J-N}$. The presence of J in the phase serves to eliminate complex quantities when N is half-integral.

We can now define a double tensor $\mathbf{D}^{(JJ)}$ through the equation

$$D_{M,-N}{}^{(JJ)} = (-1)^{J-N}(2J+1)^{1/2}\mathfrak{D}_{MN}{}^J(\omega)^*. \tag{1.27}$$

The factor $(2J+1)^{1/2}$ simplifies various calculations, as we shall see later on. It is understood that $\mathbf{D}^{(JJ)}$ refers to a specific value of the Euler triad ω.

1.7 COUPLING

The coupling of two angular momenta $\mathbf{A}$ and $\mathbf{B}$ to form a third, $\mathbf{C}$, is familiar to all spectroscopists. It is supposed that $\mathbf{A}$ and $\mathbf{B}$ refer to different physical systems, or else to independent parts of the same system. This guarantees that $\mathbf{A}$ and $\mathbf{B}$ commute, thereby ensuring that $\mathbf{C}$, which is defined by $\mathbf{C} = \mathbf{A} + \mathbf{B}$, satisfies the commutation relations (1.2). To simplify the notation as much as possible, let us write the eigenvalues of $\mathbf{A}^2$ and A_z as $A(A+1)$ and a respectively. Two types of ket are now available for describing the combined scheme. We can choose either the uncoupled form $| Aa, Bb\rangle$, or the coupled form $| Cc\rangle$. For a given A and B, there are $(2A+1)(2B+1)$ kets of the first kind; and if we use the fact that C can run with integral steps from $A + B$ down to $| A - B |$, it can be easily checked that there are an equal number of coupled kets. The unitary transformation that connects the two descriptions can be written

$$| Cc\rangle = \sum_{a,b} (Aa, Bb \mid Cc) \mid Aa, Bb\rangle. \tag{1.28}$$

The coefficients $(Aa, Bb \mid Cc)$ are the celebrated Clebsch–Gordan (CG) coefficients. Phases can be chosen so that they are all real. By operating on both sides of Eq. (1.28) with C_z $(= A_z + B_z)$, we at once see that $a + b = c$, so the sum effectively runs over a single index. Owing to the unitarity of Eq. (1.28), we have

$$\sum_{a,b} (Cc \mid Aa, Bb)(Aa, Bb \mid C'c') = \delta(c, c')\delta(C, C'), \tag{1.29}$$

$$\sum_{C,c} (Aa, Bb \mid Cc)(Cc \mid Aa', Bb') = \delta(a, a')\delta(b, b'). \tag{1.30}$$

It would be out of place in this introduction to go deeply into the properties of the CG coefficients. Nowadays they are often replaced by the 3-j symbol [8], which is defined by

$$\begin{pmatrix} A & B & C \\ a & b & c \end{pmatrix} = (-1)^{A-B-c}(2C+1)^{-1/2}(Aa, Bb \mid C\ -c). \tag{1.31}$$

The 3-j symbol exhibits the symmetries of the CG coefficient in a particularly transparent form. Even permutations of the columns leave the numerical value of the symbol unchanged, while odd permutations of the columns or a reversal in sign of the entries of the lower row produce a phase factor $(-1)^{A+B+C}$. Over the years, many tabulations of the CG coefficients and 3-j symbols have been made. The numerical tables of Rotenberg *et al.* [9] are some of the most extensive. Edmonds [1] gives the algebraic forms of a number of the more simple 3-j symbols. Several of these are assembled in Appendix I. A method for finding an explicit expression for the general CG coefficient is outlined in Problem 1.3.

To indicate that A and B are coupled to C, the ket $|\,Cc\rangle$ is often more completely written as $|\,(AB)Cc\rangle$. The arrangement of symbols is of some significance, since the explicit form for the CG coefficient appearing in Eq. (1.28) permits the interchange $Aa \leftrightarrow Bb$ only when the phase $(-1)^{A+B-C}$ is included. So

$$|\,(AB)Cc\rangle = (-1)^{A+B-C}\,|\,(BA)Cc\rangle. \tag{1.32}$$

Tensor operators, like kets, can be coupled. Thus we may write

$$(\mathbf{T}^{(A)}\mathbf{U}^{(B)})_c{}^{(C)} = \sum_{a,b} (Aa, Bb \mid Cc)\, T_a{}^{(A)} U_b{}^{(B)}. \tag{1.33}$$

It is traditional [7] to define a scalar product by the equation

$$(\mathbf{T}^{(A)}\cdot\mathbf{U}^{(A)}) = \sum_a (-1)^a T_a{}^{(A)} U_{-a}{}^{(A)}.$$

From Edmonds' text, we at once find that

$$(Aa, A\ -a \mid 00) = (-1)^{-A+a}(2A+1)^{-1/2}, \tag{1.34}$$

so that

$$(\mathbf{T}^{(A)}\cdot\mathbf{U}^{(A)}) = (-1)^A(2A+1)^{1/2}(\mathbf{T}^{(A)}\mathbf{U}^{(A)})^{(0)}. \tag{1.35}$$

The familiar vector product $\mathbf{T} \times \mathbf{U}$ formed from two vectors $\mathbf{T}$ and $\mathbf{U}$ must be proportional to $(\mathbf{T}^{(1)}\mathbf{U}^{(1)})^{(1)}$. If we decide to find the proportionality factor by equating the z component of the former to the $q = 0$ component of the latter, we need the CG coefficients

$$(11, 1-1 \mid 10) = -(1-1, 11 \mid 10) = \sqrt{\tfrac{1}{2}},$$
$$(10, 10 \mid 10) = 0,$$

as well as the relations

$$T_{\pm 1}^{(1)} = \mp(2)^{-1/2}(T_x \pm iT_y), \qquad T_0^{(1)} = T_z, \tag{1.36}$$

which parallel Eqs. (1.19). We find that

$$\mathbf{T} \times \mathbf{U} = -i\sqrt{2}\,(\mathbf{T}^{(1)}\mathbf{U}^{(1)})^{(1)}. \tag{1.37}$$

Double tensors can be coupled in an analogous way to single-rank tensors:

$$(\mathbf{T}^{(AB)}\mathbf{U}^{(CD)})_{ef}^{(EF)} = \sum_{abcd} (Aa, Cc \mid Ee)(Bb, Dd \mid Ff)\, T_{ab}^{(AB)} U_{cd}^{(CD)}.$$

To illustrate this, we consider again the two vectors $\boldsymbol{l}$ and $\boldsymbol{\lambda}$ defined in Eqs. (1.22) and (1.23). The double tensors, $\mathbf{r}^{(01)}$ and $\mathbf{r}^{(10)}$, can be readily constructed from the coordinates of position:

$$r_{00}^{(01)} = \zeta, \qquad r_{0\pm1}^{(01)} = \mp\sqrt{\tfrac{1}{2}}\,(\xi \pm i\eta),$$
$$r_{00}^{(10)} = z, \qquad r_{\pm10}^{(10)} = \mp\sqrt{\tfrac{1}{2}}\,(x \pm iy).$$

From Eq. (1.17), we know that we can write

$$\begin{aligned} r_{0q}^{(01)} &= \sum_{q'} \mathfrak{D}_{q'q}^{1}(\omega)\, r_{q'0}^{(10)} \\ &= 3^{-1/2} \sum_{q'} (-1)^{1-q'} D_{-q'q}^{(11)} r_{q'0}^{(10)} \end{aligned}$$

by Eqs. (1.14) and (1.27). Using

$$(1-q', 1q' \mid 00) = 3^{-1/2}(-1)^{-1+q'}, \qquad (1q, 00 \mid 1q) = 1,$$

we get

$$\mathbf{r}^{(01)} = (\mathbf{D}^{(11)}\mathbf{r}^{(10)})^{(01)}. \tag{1.38}$$

This is a statement of Eqs. (1.10) in tensorial form. The inverse relation can be equally well derived:

$$\mathbf{r}^{(10)} = (\mathbf{D}^{(11)}\mathbf{r}^{(01)})^{(10)}. \tag{1.39}$$

The remarkable compression of Eqs. (1.38)–(1.39) illustrates the power of the tensor method. A useful convention is to reserve the same descriptive letter ($\mathbf{r}$ in the present example) for pairs of tensors connected by the analogs of Eqs. (1.38)–(1.39). That is, $(\mathbf{D}^{(kk)}\mathbf{T}^{(k0)})^{(0k)}$ is denoted by $\mathbf{T}^{(0k)}$; similarly, $(\mathbf{D}^{(kk)}\mathbf{U}^{(0k)})^{(k0)}$ is denoted by $\mathbf{U}^{(k0)}$.

1.8 INTEGRALS OVER THE ROTATION MATRICES

In addition to coupling vectors such as $\mathbf{r}^{(10)}$ or $\mathbf{r}^{(01)}$ to $\mathbf{D}^{(11)}$, we can couple the $\mathbf{D}$ tensors among themselves. As a preliminary to doing this, we define, for an angular momentum vector $\mathbf{A}$, the following operator:

$$D_A(\omega) = \exp(-i\alpha A_z)\exp(-i\beta A_y)\exp(-i\gamma A_z).$$

If $\mathbf{A}$ commutes with another angular momentum vector $\mathbf{B}$, we can commute the exponentials of $D_A(\omega)$ with those of $D_B(\omega)$, thereby obtaining the result

$$D_A(\omega)D_B(\omega) = D_C(\omega), \tag{1.40}$$

where $\mathbf{C} = \mathbf{A} + \mathbf{B}$. We now set both sides of Eq. (1.40) between $\langle Aa, Bb \mid$ and $\mid Aa', Bb' \rangle$. With the aid of Eq. (1.28), we arrive at the equation

$$\mathfrak{D}_{aa'}{}^{A}(\omega)\mathfrak{D}_{bb'}{}^{B}(\omega) = \sum_{Ccc'} (Aa, Bb \mid Cc)(Aa', Bb' \mid Cc')\mathfrak{D}_{cc'}{}^{C}(\omega). \tag{1.41}$$

This can be thrown into a tensorial form by converting the $\mathfrak{D}$ functions to D tensors and coupling the two spaces to ranks T and T' respectively. The result is

$$(\mathbf{D}^{(AA)}\mathbf{D}^{(BB)})^{(TT')} = \delta(T, T')\sqrt{\frac{(2A+1)(2B+1)}{(2T+1)}}\,\mathbf{D}^{(TT)}. \tag{1.42}$$

We have not yet considered any kind of integration with respect to the Euler angles ω. To evaluate

$$\int \mathfrak{D}_{aa'}{}^{A}(\omega)\mathfrak{D}_{bb'}{}^{B}(\omega)\,d\omega,$$

we must first decide on the form for $d\omega$. The natural choice is

$$d\omega = d\alpha \sin\beta\, d\beta\, d\gamma,$$

for this ensures that all orientations of a frame F' with respect to a fixed frame F are given equal weight as the angles α, β, and γ (that relate F' to F) run over their allotted ranges, namely

$$0 \le \alpha \le 2\pi, \qquad 0 \le \beta \le \pi, \qquad 0 \le \gamma \le 2\pi. \tag{1.43}$$

At this point, we can simply appeal to the relation that exists between the $\mathfrak{D}$ functions and the Jacobi polynomials (as does Edmonds [1]). However, it is possible to avoid this approach by making use of Eq. (1.41) and the result of Problem 1.4. The integration of the single function $\mathfrak{D}_{cc'}{}^{C}(\omega)$ over γ yields zero unless $c' = 0$; and, for this case, the $\mathfrak{D}$ function is proportional to the spherical harmonic $Y_{Cc}(\beta, \alpha)$. The integration over β and α yields a null result unless $C = c = 0$. So, using the fact that $\mathfrak{D}_{00}{}^{0} = C_0{}^{(0)} = 1$, we have

$$\int \mathfrak{D}_{aa'}{}^{A}(\omega)\mathfrak{D}_{bb'}{}^{B}(\omega)\, d\omega = (Aa, Bb \mid 00)(Aa', Bb' \mid 00) \int d\omega$$

$$= 8\pi^2 \delta(A, B)\delta(a + b, 0)\delta(a' + b', 0)/(2A + 1).$$

In getting this result, Eq. (1.34) has been used to evaluate the CG coefficients. An alternative form is

$$\int \mathfrak{D}_{MN}{}^{J}(\omega)^* \mathfrak{D}_{M'N'}{}^{J'}(\omega)\, d\omega = \frac{8\pi^2}{2J+1}\delta(J, J')\delta(M, M')\delta(N, N'). \quad (1.44)$$

The repeated use of Eq. (1.41) allows any multiple product of $\mathfrak{D}$ functions to be reduced to a manageable form. For example, we can show without difficulty that

$$\int \mathfrak{D}_{aa'}{}^{A}(\omega)\mathfrak{D}_{bb'}{}^{B}(\omega)\mathfrak{D}_{cc'}{}^{C}(\omega)\, d\omega = 8\pi^2 \begin{pmatrix} A & B & C \\ a & b & c \end{pmatrix}\begin{pmatrix} A & B & C \\ a' & b' & c' \end{pmatrix}. \quad (1.45)$$

It is interesting to note that Eq. (1.44) has an analog in the theory of finite groups. Consider an irreducible unitary matrix representation Γ_P of a finite group G. Let the entries in the matrix representing a particular group element σ be $p_{ij}(\sigma)$, where i and j label the rows and columns. From Schur's lemma, it can be shown [10] that

$$\sum_{\sigma} p_{ij}(\sigma)^* p_{lk}(\sigma) = g\delta(i, l)\delta(j, k)/n_P, \quad (1.46)$$

where n_P is the number of rows (or columns) in the matrices p_{ij}, and g is the number of elements comprising G. Corresponding to a summation of the finite group, we have an integration over the group elements of $R(3)$ in Eq. (1.44). That our matrices $\mathfrak{D}_{MN}{}^{J}$ form a matrix representation for the rotation group $R(3)$ can be established by noting that, if

$$D(\omega) = D(\omega_1)D(\omega_2), \quad (1.47)$$

then the insertion of both sides of this equation between $\langle JM \mid$ and $\mid JN \rangle$ yields

$$\mathfrak{D}_{MN}{}^{J}(\omega) = \sum_{T} \mathfrak{D}_{MT}{}^{J}(\omega_1)\mathfrak{D}_{TN}{}^{J}(\omega_2). \quad (1.48)$$

The sum on the right-hand side of this equation corresponds exactly to matrix multiplication. Our labels MN and $M'N'$ thus correspond to ij and lk, and Eq. (1.46) is the analog of Eq. (1.44) for $J = J'$. (A similar analog exists for $J \neq J'$.) This analogy can be strengthened to provide an alternative derivation for Eq. (1.44).

1.9 THE WIGNER–ECKART THEOREM

We are now in a position to derive the central theorem of spectroscopy. It concerns the factorization of the matrix elements

$$\langle JM \mid T_q^{(k)} \mid J'M' \rangle,$$

where the rank and component of the operator, as well as the quantum numbers in the bra and ket, are defined with respect to a single angular momentum vector $\mathbf{J}$. The coupling of states and the coupling of operators have been described in Section 1.7. It is equally easy to couple an operator to a state. For example, we can write

$$T_q^{(k)} \mid J'M' \rangle = \sum_{KN} (KN \mid kq, J'M') \mid KN \rangle, \qquad (1.49)$$

where the unnormalized kets $\mid KN \rangle$ are given by

$$\mid KN \rangle = (\mathbf{T}^{(k)} \mid J' \rangle)_N{}^{(K)}. \qquad (1.50)$$

The symbol $\mid J' \rangle$ stands for the $2J' + 1$ kets $\mid J'M' \rangle$ for which $-J' \leq M' \leq J'$ in just the same way as $\mathbf{T}^{(k)}$ stands for the $2k + 1$ operators $T_q^{(k)}$ for which $-k \leq q \leq k$. From Eq. (1.49), we at once get

$$\langle JM \mid T_q^{(k)} \mid J'M' \rangle = \sum_{KN} \langle JM \mid KN \rangle (KN \mid kq, J'M').$$

Now

$$\langle JM \mid KN \rangle = \langle JM \mid D(\omega)^{-1} D(\omega) \mid KN \rangle$$
$$= \sum_{M'N'} \mathfrak{D}_{M'M}{}^{J}(\omega)^* \mathfrak{D}_{N'N}{}^{K}(\omega) \langle JM' \mid KN' \rangle.$$

Multiplying both sides by $d\omega/8\pi^2$, and integrating over ω, we get

$$\langle JM \mid KN \rangle = (2J + 1)^{-1} \delta(J, K) \delta(M, N) \sum_{M'N'} \delta(M', N') \langle JM' \mid KN' \rangle$$

by means of Eq. (1.44). So $\langle JM \mid KN \rangle$ vanishes unless $K = J$ and $M = N$, and when these conditions hold, it is independent of M. Thus we can write

$$\langle JM \mid T_q^{(k)} \mid J'M' \rangle = \Xi (JM \mid kq, J'M'), \qquad (1.51)$$

where Ξ is independent of M, q, and M'. Although Eq. (1.51) is commonly referred to as the Wigner–Eckart (WE) theorem, it is more properly thought of as a specialization of the theorem to the group $R(3)$. The derivation given above is similar in principle (though not in detail) to that of Wigner himself [10]. O'Raifeartaigh has generalized Wigner's method to all compact simple Lie groups [11].

When the CG coefficient is replaced by a 3-j symbol, it is usual to write

$$\Xi = (-1)^{J-J'+k}(2J+1)^{-1/2}(J \,\|\, T^{(k)} \,\|\, J'),$$

so that the WE theorem runs

$$\langle JM \mid T_q^{(k)} \mid J'M' \rangle = (-1)^{J-M}\begin{pmatrix} J & k & J' \\ -M & q & M' \end{pmatrix}(J \,\|\, T^{(k)} \,\|\, J'). \quad (1.52)$$

This is the form most widely used at present. The quantity $(J \,\|\, T^{(k)} \,\|\, J')$ is called a *reduced* matrix element. It is important to recognize that a reduced matrix element is merely a number, and only in a very special sense is it some combination of a bra, an operator, and a ket. The fact that it looks like a matrix element is only a notational convenience. Additional quantum numbers that may be necessary to more closely define the bra and the ket of the original matrix element can be immediately carried over into the reduced matrix element, which is then ready for further analysis.

The importance of the WE theorem can hardly be overstated. Our attitude to physical systems is often determined by it, although we may not always be conscious of the fact. For example, its application to nuclear eigenfunctions allows the atomic and molecular physicist to replace the nuclear magnetic moment $\boldsymbol{\mu}_N$ by a number times $\mathbf{I}$, the nuclear spin. This is possible because both $\boldsymbol{\mu}_N$ and $\mathbf{I}$ are tensors of unit rank with respect to $\mathbf{I}$, and hence their matrix elements, taken between states of the same I, differ only in the reduced part. The WE theorem permits us to take full advantage of any symmetry that may be inherent in a problem; in other words, to extract those parts that are essentially geometric in character.

1.10 REDUCED MATRIX ELEMENTS

The straightforward way to find a reduced matrix element is to apply the WE theorem to the simplest special case involving the quantities of interest. A particularly common reduced matrix element involves an angular momentum vector $\mathbf{A}$ set between its own eigenstates. To evaluate it, we compare the equation

$$\langle Aa \mid A_z \mid Aa \rangle = a$$

with Eq. (1.52);

$$\langle Aa \mid A_0^{(1)} \mid Aa \rangle = (-1)^{A-a} \begin{pmatrix} A & 1 & A \\ -a & 0 & a \end{pmatrix} (A \,\|\, A \,\|\, A).$$

Since $A_z = A_0^{(1)}$, we can equate the right-hand sides of these two equations. On putting in the explicit form for the 3-j symbol (see Appendix I), we get

$$(A \,\|\, A \,\|\, A) = [A(A+1)(2A+1)]^{1/2}. \tag{1.53}$$

As expected, this is independent of a.

Another common reduced matrix element arises when the tensor $\mathbf{C}^{(k)}$ is set between bras $\langle lm \mid$ and kets $\mid l'm' \rangle$ that denote the spherical harmonics Y_{lm}^* and $Y_{l'm'}$ respectively. Setting $a' = b' = c' = 0$ in Eq. (1.45), and using the result of Problem 1.4, we have

$$\iint Y_{Aa}(\theta, \phi) Y_{Bb}(\theta, \phi) Y_{Cc}(\theta, \phi) \sin\theta \, d\theta \, d\phi$$

$$= \sqrt{\frac{[A, B, C]}{4\pi}} \begin{pmatrix} A & B & C \\ a & b & c \end{pmatrix} \begin{pmatrix} A & B & C \\ 0 & 0 & 0 \end{pmatrix}, \tag{1.54}$$

where the abbreviation [12]

$$[P, Q, \ldots, T] = [(2P+1)(2Q+1)\cdots(2T+1)] \tag{1.55}$$

has been used. Now, from Problem 1.4 and Eq. (1.14),

$$\begin{aligned} Y_{Aa}(\theta, \phi) &= [(2k+1)/4\pi]^{1/2} \mathfrak{D}_{a0}^{A}(\phi\theta\gamma)^* \\ &= [(2k+1)/4\pi]^{1/2} (-1)^a \mathfrak{D}_{-a0}^{A}(\phi\theta\gamma) \\ &= (-1)^a Y_{A-a}^*(\theta, \phi). \end{aligned} \tag{1.56}$$

Setting

$$A = l, \quad a = m, \quad B = k, \quad b = q, \quad C = l', \quad c = m'$$

in Eq. (1.54), we get

$$\langle lm \mid C_q^{(k)} \mid l'm' \rangle = [l, l']^{1/2} (-1)^m \begin{pmatrix} l & k & l' \\ -m & q & m \end{pmatrix} \begin{pmatrix} l & k & l' \\ 0 & 0 & 0 \end{pmatrix},$$

from which we can immediately deduce that

$$(l \,\|\, C^{(k)} \,\|\, l') = (-1)^l [l, l']^{1/2} \begin{pmatrix} l & k & l' \\ 0 & 0 & 0 \end{pmatrix}. \tag{1.57}$$

For the 3-j symbol not to vanish, $l + k + l'$ must be even.

In the case of double tensors, the WE theorem runs

$$\langle J_1M_1, J_2M_2 \mid T_{q_1q_2}{}^{(k_1k_2)} \mid J_1'M_1', J_2'M_2' \rangle$$
$$= (-1)^{J_1+J_2-M_1-M_2} \begin{pmatrix} J_1 & k_1 & J_1' \\ -M_1 & q_1 & M_1' \end{pmatrix} \begin{pmatrix} J_2 & k_2 & J_2' \\ -M_2 & q_2 & M_2' \end{pmatrix}$$
$$\times (J_1J_2 \| T^{(k_1k_2)} \| J_1'J_2'). \quad (1.58)$$

As in Eq. (1.52), we emphasize the fact that a reduced matrix element is a scalar by printing T in italics. This is in accord with Racah's original notation [7], though a corruption of comparatively recent origin, which is now quite widespread, puts T in boldface. Although harmless in itself, it can lead to errors of a rather subtle kind when double tensors are used. For if $\mathbf{T}^{(0k)} = \mathbf{U}^{(k)}$, we can easily show from Eqs. (1.52) and (1.58) that

$$(J_1J_2 \| T^{(0k)} \| J_1J_2') = (2J_1+1)^{1/2}(J_2 \| U^{(k)} \| J_2').$$

When boldface symbols are used in the reduced matrix elements, it is uncomfortably easy to overlook the factor $(2J_1+1)^{1/2}$, particularly if the symbols J_1 in the reduced matrix element are suppressed.

Equation (1.50) for a coupled operator and ket provides us with a useful way of interpreting a reduced matrix element. The analysis that leads up to Eqs. (1.51)–(1.52) yields

$$\langle JM \mid (\mathbf{T}^{(k)} \mid J' \rangle)_N{}^{(K)} = \delta(J, K)\delta(M, N)\Xi$$
$$= \delta(J, K)\delta(M, N)(-1)^{J-J'+k}$$
$$\times (2J+1)^{-1/2}(J \| T^{(k)} \| J'). \quad (1.59)$$

1.11 THE 9-j SYMBOL

As a preliminary to a study of more complicated matrix elements than those appearing in the previous section, we consider the transformation from LS to jj coupling in an atomic system comprising two electrons. The two types of ket are

$$\mid (s_1s_2)S(l_1l_2)LJM \rangle$$

and

$$\mid (s_1l_1)j_1(s_2l_2)j_2JM \rangle.$$

The couplings implied by these kets is made clear in Fig. 1.1. The coupling coefficient that intervenes when one of these kets is expanded as a sum over those of the other type is independent of M and can be written as

$$((s_1s_2)S(l_1l_2)LJ \mid (s_1l_1)j_1(s_2l_2)j_2J).$$

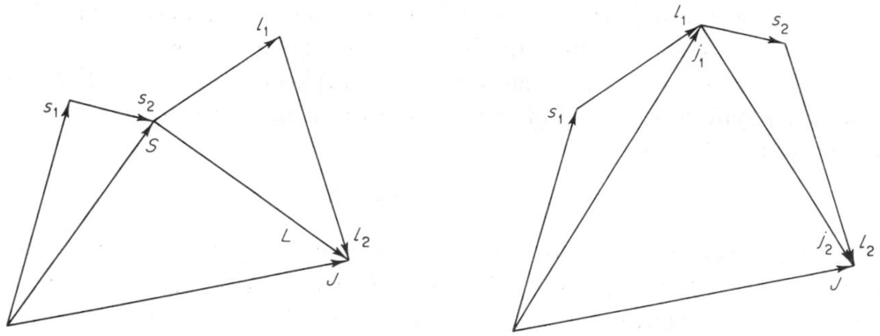

Fig. 1.1. Two alternative coupling schemes. That for LS coupling is shown on the left, that for jj coupling on the right. The symbols against the vectors can be interpreted in two ways. They can be taken to represent the angular momentum vectors themselves, in which case the couplings exhibited in the diagrams imply corresponding equations for the components of the vectors, e. g., $(s_1)_x + (s_2)_x = S_x$. Alternatively, they can be taken to represent quantum numbers. In this case, the diagrams determine the acceptable coupling schemes simply by interpreting the length of each vector as the attached quantum number. The overlap between the states represented by these two diagrams is given directly in terms of a 9-j symbol by the expression (1.60).

The running indices are either S and L or j_1 and j_2. The 9-j symbol is defined by equating this coupling coefficient to

$$[S, L, j_1, j_2]^{1/2} \begin{Bmatrix} s_1 & s_2 & S \\ l_1 & l_2 & L \\ j_1 & j_2 & J \end{Bmatrix} . \tag{1.60}$$

For the 9-j symbol not to vanish, the three members of each row and each column must satisfy the triangular conditions: that is, it must be possible to form a triangle whose sides are of the lengths specified by the quantum numbers. This is merely a statement that all quantum numbers fall within the ranges permitted by the couplings. Various tabulations [13] of the 9-j symbols have been made, but the limited range of arguments that it is feasible to have printed makes it usually more convenient to depend on a computer program.

The first and second columns of the 9-j symbol can be reversed by interchanging the suffixes 1 and 2. The reordering of the couplings implied by this can be revoked by repeated use of Eq. (1.32). The phase factor introduced in this operation is

$$(-1)^{s_1+s_2-S+l_1+l_2-L+j_1+j_2-J},$$

and since $S + L + J$ must always be integral, the phase factor reduces to $(-1)^R$, where R is the sum of all the arguments that appear in the 9-j symbol. This result in fact holds good for the interchange of any two columns or of any two rows, as well as for a reflection in either diagonal.

The 9-j symbol is useful in the evaluation of matrix elements whenever the bare description

$$\langle JM \mid T_q^{(k)} \mid J'M' \rangle$$

can be augmented by more information concerning the way in which J is constructed. For example, if $\mathbf{J} = \mathbf{A} + \mathbf{B}$, where $\mathbf{A}$ and $\mathbf{B}$ are commuting angular momenta, we can examine the tensorial properties of the states and the operators with respect to these angular momenta. This leads us to ask for the properties of the reduced matrix elements

$$((AB)C \parallel (\mathbf{U}^{(A')}\mathbf{V}^{(B')})^{(C')} \parallel (A''B'')C''). \tag{1.61}$$

To unify the notation, the substitutions $J = C$, $k = C'$, and $J' = C''$ have been made.

If we use Eq. (1.59), the reduced matrix element (1.61) becomes

$$(-1)^{C-C''+C'}(2C+1)^{1/2}\langle (AB)Cc \mid [(\mathbf{U}^{(A')}\mathbf{V}^{(B')})^{(C')} \mid (A''B'')C''\rangle]_c^{(C)}.$$

We now recouple the quantity in brackets, getting

$$(-1)^{C+C'-C''}(2C+1)^{1/2}\sum_{EF}\langle (AB)Cc \mid [(\mathbf{U}^{(A')} \mid A''\rangle)^{(E)}(\mathbf{V}^{(B')} \mid B''\rangle)^{(F)}]_c^{(C)}$$

$$\times\ ((A'A'')E(B'B'')FC \mid (A'B')C'(A''B'')C''C)$$

$$= (-1)^{C+C'-C''}(2C+1)^{1/2}\sum_{EFabef}(Cc \mid Aa, Bb)(Ee, Ff \mid Cc)$$

$$\times\ ((A'A'')E(B'B'')FC \mid (A'B')C'(A''B'')C''C)$$

$$\times\ \langle Aa \mid (\mathbf{U}^{(A')} \mid A''\rangle)_e^{(E)}\langle Bb \mid (\mathbf{V}^{(B')} \mid B''\rangle)_f^{(F)}.$$

The last two factors can be converted into reduced matrix elements with the aid of Eq. (1.59). The delta functions in that equation yield $A = E$, $a = e$, $B = F$, and $b = f$. The sum over $(abef)$ can now be performed by using Eq. (1.29). On replacing the recoupling coefficient by a 9-j symbol, we arrive at the final equation

$$((AB)C \parallel (\mathbf{U}^{(A')}\mathbf{V}^{(B')})^{(C')} \parallel (A''B'')C'')$$

$$= [C, C', C'']^{1/2}\begin{Bmatrix} A & A'' & A' \\ B & B'' & B' \\ C & C'' & C' \end{Bmatrix}(A \parallel U^{(A')} \parallel A'')(B \parallel V^{(B')} \parallel B''). \tag{1.62}$$

In deriving this result, we have assumed that the ket $| A'' \rangle$ and the tensor $\mathbf{V}^{(B')}$ commute. This condition holds for all the uses we shall put this formula to. The more general form, which involves a sum over the additional classificatory symbol γ'', is given by Edmonds [1].

1.12 THE 6-j SYMBOL

A number of simplifications take place when one of the arguments of a 9-j symbol is zero. Without loss of generality, we can suppose the null entry occurs in the middle of the first column. In this case we write

$$((AB)C, D, E \mid A, (BD)F, E) = [A, C, D, F]^{1/2} \begin{Bmatrix} A & B & C \\ 0 & D & D \\ A & F & E \end{Bmatrix}$$

$$= (-1)^{A+B+D+E}[C, F]^{1/2} \begin{Bmatrix} A & B & C \\ D & E & F \end{Bmatrix}. \tag{1.63}$$

These equations define the 6-j symbol as a recoupling coefficient and also as a special case of a 9-j symbol. From the symmetries of the 9-j symbol, we can show that columns of the 6-j symbol can be interchanged without affecting its numerical value; furthermore, the upper and lower entries in a pair of columns can be simultaneously interchanged. These properties make it much simpler to handle the 6-j symbol than Racah's original W function [7]. Extensive tables of the 6-j symbol are available [9], as well as algebraic expressions for some of the more elementary ones. A limited selection of these is given in Appendix I.

The general expression for the 6-j symbol is not easy to obtain. The obvious method is to completely uncouple the left and right parts of the recoupling coefficient of Eq. (1.63) and then use

$$(Aa, Bb, Cc \mid Aa', Bb', Cc') = \delta(a, a')\delta(b, b')\delta(c, c').$$

Since there are four uncouplings to perform, the 6-j symbol is expressible as a sum over a product of four 3-j symbols, for which explicit expressions are available. With considerable ingenuity, the sums can be reduced to only one. This, in essence, is Racah's method. A different approach, in which elaborate and extraordinarily subtle use is made of generating functions, has been devised by Schwinger [14]. Although this derivation of the formula for a 6-j symbol is certainly the more elegant one, a considerable set of preliminary lemmas is required. We content ourselves simply by a statement of the result:

$$\begin{Bmatrix} A & B & C \\ D & E & F \end{Bmatrix}$$
$$= \Delta(ABC)\,\Delta(AEF)\,\Delta(DBF)\,\Delta(DEC)$$
$$\times \sum_z \frac{(-1)^z(z+1)!}{\begin{aligned}&[(z-A-B-C)!(z-A-E-F)!(z-D-B-F)!\\ &\cdot(z-D-E-C)!(A+B+D+E-z)!\\ &\cdot(A+C+D+F-z)!(B+C+E+F-z)!]\end{aligned}}, \tag{1.64}$$

where

$$\Delta(XYZ) = \left\{\frac{(X+Y-Z)!(X+Z-Y)!(Y+Z-X)!}{(X+Y+Z+1)!}\right\}^{1/2}.$$

Equation (1.62) yields three important special cases on setting either A', B', or C' to zero. When $A' = 0$, the 9-j symbol vanishes unless $A'' = A$. Putting $\mathbf{U}^{(0)} = 1$, we get

$$((AB)C \,\|\, V^{(B')} \,\|\, (AB'')C'')$$
$$= (-1)^{A+B''+B'+C}[C, C'']^{1/2} \begin{Bmatrix} B & B' & B'' \\ C'' & A & C \end{Bmatrix} (B \,\|\, V^{(B')} \,\|\, B''). \tag{1.65}$$

This equation is appropriate to a tensor $\mathbf{V}^{(B')}$ that commutes with $\mathbf{A}$. It is sometimes said that $\mathbf{V}^{(B')}$ acts on the second part of the coupled system. Similarly, when $B' = 0$, the corresponding result is

$$((AB)C \,\|\, U^{(A')} \,\|\, (A''B)C'')$$
$$= (-1)^{A+A'+B+C''}[C, C'']^{1/2} \begin{Bmatrix} A & A' & A'' \\ C'' & B & C \end{Bmatrix} (A \,\|\, U^{(A')} \,\|\, A''). \tag{1.66}$$

This equation applies to a tensor $\mathbf{U}^{(A')}$ that commutes with $\mathbf{B}$ and thus acts on the first part of a coupled system. Finally, when $C' = 0$, we get, with the aid of Eq. (1.35), a formula for the reduced matrix element of a scalar product. It is usually more convenient to have the matrix element itself rather than its reduced part. In the notation of Section 1.7,

$$\langle (AB)Cc \mid (\mathbf{U}^{(A')}\cdot\mathbf{V}^{(A')}) \mid (A''B'')C''c'' \rangle$$
$$= (-1)^{A''+B+C}\delta(C, C'')\delta(c, c'') \begin{Bmatrix} A & A' & A'' \\ B'' & C & B \end{Bmatrix}$$
$$\times (A \,\|\, U^{(A')} \,\|\, A'')(B \,\|\, V^{(A')} \,\|\, B''). \tag{1.67}$$

This result holds good when $\mathbf{U}^{(A')}$ and $\mathbf{V}^{(A')}$ act on the first and second parts respectively of a coupled system; that is, $\mathbf{U}^{(A')}$ commutes with $\mathbf{B}$, and $\mathbf{V}^{(A')}$ commutes with $\mathbf{A}$.

If A or B (and, simultaneously, A'' or B'') are themselves comprised of coupled parts, the equations above can be reapplied to the reduced matrix elements on their right-hand sides, thereby giving a product of 6-j symbols. The process ends when the reduced matrix elements are in such a primitive form that their values can be written down. Double tensors can be treated in precisely the same way. We have only to replace every bra and ket by ones in which two distinct spaces appear, and then double up all the numerical coefficients (including the 6-j symbols) to allow for the existence of the two spaces. The extension of the simple WE theorem to the form appropriate for double tensors (as given in Eq. 1.58) should make the procedure clear.

PROBLEMS

1.1 A coordinate frame F is brought to F' by a rotation of α about the z axis, and then to F'' by a rotation of β about the axis y' (the y axis of F'). Prove that

$$\exp(-i\alpha J_z)\, J_y \exp(i\alpha J_z) = J_{y'}$$

and hence that

$$\exp(-i\beta J_{y'}) \exp(-i\alpha J_z) = \exp(-i\alpha J_z) \exp(-i\beta J_y).$$

The frame F'' is brought to F''' by a rotation of γ about the axis z'' (the z axis of F''). Prove that

$$\exp(-i\gamma J_{z''}) \exp(-i\beta J_{y'}) = \exp(-i\beta J_{y'}) \exp(-i\gamma J_z),$$

and combine the results to obtain

$$\begin{aligned}&\exp(-i\alpha J_z) \exp(-i\beta J_y) \exp(-i\gamma J_z)\\ &\quad= \exp(-i\gamma J_{z''}) \exp(-i\beta J_{y'}) \exp(-i\alpha J_z).\end{aligned}$$

1.2 Prove that the commutation relations between the components of $\mathbf{J}$, as well as the action of these operators on $|JM\rangle$, remain valid in a representation defined by the following replacements:

$$J_+ \to -\nu\frac{\partial}{\partial\mu}, \qquad J_- \to -\mu\frac{\partial}{\partial\nu}, \qquad J_z \to -\frac{1}{2}\left(\mu\frac{\partial}{\partial\mu} - \nu\frac{\partial}{\partial\nu}\right),$$

$$|JM\rangle \to (-1)^{J-M}[(J-M)!(J+M)!]^{-1/2}\mu^{J-M}\nu^{J+M}.$$

Make the substitutions $\mu = \rho \sin \varphi$, $\nu = \rho \cos \varphi$, and find J_y in terms of ρ and φ. Prove

$$\exp(\tfrac{1}{2}\beta\partial/\partial\varphi)(\sin^a \varphi \cos^b \varphi) = \sin^a(\varphi + \tfrac{1}{2}\beta) \cos^b(\varphi + \tfrac{1}{2}\beta).$$

Expand the quantities on the right-hand side of this equation and thereby derive Eq. (1.11) (see also Rose [15]).

1.3 Two commuting angular momenta **A** and **B** are coupled to form a third, **C**. Prove that the expression

$$\sum_{a,b} (-1)^{A-a}\delta(a+b, C) \left(\frac{(A+a)!(B+b)!}{(A-a)!(B-b)!}\right)^{1/2} | Aa, Bb\rangle$$

gives zero when acted on by C_+. Deduce that it must correspond to a multiple of the ket $| Cc\rangle$ for which $c = C$. Prove that, to normalize it, the factor

$$\left[\frac{(C+A-B)!(C+B-A)!(C+A+B+1)!}{(2C+1)!(A+B-C)!}\right]^{1/2}$$

must be prefixed to it. Operate on $| CC\rangle$ with $(C_-)^{C-c}$, and thereby obtain the result

$$\begin{aligned}(Aa, Bb \mid Cc) &= \delta(a+b, c) \\ &\times \left(\frac{(2C+1)(A+B-C)!(A-a)!(B-b)!(C+c)!(C-c)!}{(A+B+C+1)!(C+A-B)!(C+B-A)!(A+a)!(B+b)!}\right)^{1/2} \\ &\times \sum_x (-1)^{A-a-x} \frac{(A+a+x)!(B+C-a-x)!}{x!(C-c-x)!(A-a-x)!(B-C+a+x)!}.\end{aligned}$$

(The details of this derivation can be found elsewhere [16].)

1.4 By writing

$$\frac{d^{l+m}}{d\mu^{l+m}}(\mu^2-1)^l = \frac{d^{l+m}}{d\mu^{l+m}}\{(\mu-1)^l(\mu+1)^l\},$$

prove

$$\begin{aligned}P_l^m(\mu) &= \frac{1}{2^l}\sum_t (-1)^{t-m} \frac{l!(l+m)!}{(l-t)!(t-m)!t!(l+m-t)!} \\ &\quad \times (1+\mu)^{l-t+\frac{1}{2}m}(1-\mu)^{t-\frac{1}{2}m}.\end{aligned}$$

Deduce that

$$C_m^{(l)}(\beta\alpha) = (\mathfrak{D}_{m0}{}^l(\alpha\beta\gamma))^* = (-1)^m\mathfrak{D}_{0m}{}^l(-\gamma\beta-\alpha)$$

in the notation of Section 1.5.

1.5 Prove that the commutation relations (1.18) between the components of a tensor $\mathbf{T}^{(k)}$ and those of the angular momentum vector $\mathbf{J}$ can be written as

$$[J_p^{(1)}, T_q^{(k)}] = -\{k(k+1)\}^{1/2}(1p, kq \mid k\,q+p)T_{q+p}{}^{(k)}.$$

1.6 The symbol $\langle J \mid$ stands for the collection of bras whose M component is $(-1)^{J-M}\langle J, -M \mid$. Prove that

$$\sum_M \mid JM\rangle\langle JM \mid = (2J+1)^{1/2}\{\mid J\rangle\langle J \mid\}^{(0)}$$

and

$$(J \,||\, T^{(k)} \,||\, J') = (-1)^{J+k-M'}(2J'+1)^{-1/2}[\langle J \mid \mathbf{T}^{(k)}]_{M'}{}^{(J')} \mid J', -M'\rangle$$

for any M'. Derive the equation

$$(\mathbf{U}^{(k)}\mathbf{V}^{(k')})^{(K)} = \sum_{JJ'J''} (-1)^{k+k'+K+2J'}[J, J'']^{1/2} \begin{Bmatrix} k' & K & k \\ J & J' & J'' \end{Bmatrix}$$

$$\times \{(\mathbf{U}^{(k)} \mid J'\rangle)^{(J)}(\langle J' \mid \mathbf{V}^{(k')})^{(J'')}\}^{(K)},$$

and use it to obtain the result

$$(J \,||\, (\mathbf{U}^{(k)}\mathbf{V}^{(k')})^{(K)} \,||\, J'') = (-1)^{J+K+J''}(2K+1)^{1/2} \sum_{J'} \begin{Bmatrix} k' & K & k \\ J & J' & J'' \end{Bmatrix}$$

$$\times (J \,||\, U^{(k)} \,||\, J')(J' \,||\, V^{(k')} \,||\, J'').$$

(For an alternative derivation, see Edmonds [1].)

2

The Rotation Group R(4)

2.1 GROUPS

The close connection between angular momentum theory and the rotation group $R(3)$ derives from the fact that the rotation operators $D(\omega)$ are constructed in terms of the components of an angular momentum vector. Much of the analysis of the previous chapter could thus be described in the language of group theory. We have not done this to any great extent because $R(3)$ is one of the rare cases where all group-theoretical operations can be given an explicit analytical form. For example, definite expressions are available for the CG coefficients and the 6-j symbols, whereas for most other groups (such as $R(5)$), the corresponding quantities are only known for certain special cases. The privileged position of $R(3)$ can be extended, however, to one higher group, namely the rotation group in four dimensions, $R(4)$. This is particularly relevant, since two physical systems of direct interest to us, the rigid rotator and the hydrogen atom, provide examples of this group. Although a treatment of $R(4)$ could be given without drawing on the detailed apparatus of group theory, it is nevertheless very useful for us to introduce some of the more important concepts of the subject, if only to provide a working vocabulary for the analysis.

2.2 GENERATORS FOR $R(4)$

The operators $D(\omega)$ are the *elements* of $R(3)$. They are formed from the components of $\mathbf{J}$ and the Euler angles ω. The former are termed the *generators* of $R(3)$: they satisfy certain commutation relations among themselves and can be thought of as carrying out the act of rotation. The Euler angles, or *parameters*, are numbers, and determine the extent of the rotation. In generalizing $R(3)$ to $R(4)$, we note that a generator such as J_y carries out a rotation in the xz plane; it is thus more appropriate to use the notation J_{ab} and allow a and b to run over the four coordinates subject to the condition $a \neq b$. The mutual dependence of J_{ab} and J_{ba} can be formalized by writing $J_{ab} = -J_{ba}$. In this notation, the commutation relations (1.2) become

$$[J_{ab}, J_{cd}] = i\{\delta(b, d)J_{ac} - \delta(b, c)J_{ad} - \delta(a, d)J_{bc} + \delta(a, c)J_{bd}\} \qquad (2.1)$$

provided we identify $J_x = J_{yz}$, etc. In the form of Eq. (2.1), the commutation relations can be taken over immediately to $R(4)$ (or, for that matter, to any rotation group).

Let us break the six independent generators of $R(4)$ into two groups of three. We write

$$L_x = J_{23}, \qquad L_y = J_{31}, \qquad L_z = J_{12},$$

$$A_x = J_{14}, \qquad A_y = J_{24}, \qquad A_z = J_{34}. \qquad (2.2)$$

By omitting the fourth coordinate in the definition of L_x, L_y, and L_z, we obtain a three-dimensional angular momentum vector $\mathbf{L}$. We can quickly verify from Eq. (2.1) such equations as

$$[L_x, A_y] = iA_z,$$

which show that $\mathbf{A}$ is a vector with respect to $\mathbf{L}$. However, the components of $\mathbf{A}$ do not satisfy the commutation relations of an angular momentum vector. This deficiency can be corrected by defining two vectors $\mathbf{K}_1$ and $\mathbf{K}_2$ as follows:

$$\mathbf{K}_1 = \tfrac{1}{2}(\mathbf{L} + \mathbf{A}), \qquad \mathbf{K}_2 = \tfrac{1}{2}(\mathbf{L} - \mathbf{A}). \qquad (2.3)$$

It is straightforward to confirm that all components of $\mathbf{K}_1$ commute with all components of $\mathbf{K}_2$:

$$[\mathbf{K}_1, \mathbf{K}_2] = 0. \qquad (2.4)$$

Furthermore, $\mathbf{K}_1$ satisfies the commutation relations of an angular momentum vector, e.g.,

$$[K_{1x}, K_{1y}] = iK_{1z},$$

and so does $\mathbf{K}_2$. Thus we have two angular momentum vectors, $\mathbf{K}_1$ and $\mathbf{K}_2$, which can both serve as the generators for an $R(3)$ group. These two groups are independent in the sense that every generator of one commutes with every generator of the other. Moreover, the combined tally of six independent generators exhausts those of $R(4)$. The two $R(3)$ groups are said to form a *direct product,* and we write $R(4) = R(3) \times R(3)$. It follows that the properties of $R(4)$ can be built up from those of $R(3)$.

If we are working in the space of functions of the four coordinates x_a $(a = 1, 2, 3, 4)$, then the explicit form

$$J_{ab} \rightarrow \frac{1}{i}\left(x_a \frac{\partial}{\partial x_b} - x_b \frac{\partial}{\partial x_a}\right) = l_{ab} \tag{2.5}$$

is a natural choice for the generators. This is analogous to our use of $\boldsymbol{l}$ in Section 1.3 for performing the operations of $R(3)$. Since l_{ab} commutes with $x_1^2 + x_2^2 + x_3^2 + x_4^2$, we immediately see that

$$\exp(-i\theta_{ab}l_{ab})(x_1^2 + x_2^2 + x_3^2 + x_4^2)\exp(i\theta_{ab}l_{ab}) = x_1^2 + x_2^2 + x_3^2 + x_4^2,$$

so the radial distance in the four-dimensional space is left invariant, as we expect of a rotation operator. With the substitutions (2.5), all the equations of the earlier part of this section remain valid: but, in addition to these, we can also prove $\mathbf{L}\cdot\mathbf{A} = \mathbf{A}\cdot\mathbf{L} = 0$. This has a profound effect on the subsequent analysis, because we can now write

$$\mathbf{K}_1^2 = \mathbf{K}_2^2 = \tfrac{1}{4}(\mathbf{L}^2 + \mathbf{A}^2) = \mathbf{k}^2, \tag{2.6}$$

say. The operators $\mathbf{K}_1^2$ and $\mathbf{K}_2^2$ no longer have distinct eigenvalues. The irreducible representations of $R(4)$ are restricted to those that are products of two identical irreducible representations of $R(3)$. It is well known that the use of $\boldsymbol{l}$ in $R(3)$ never permits the half-integral representations to appear. The corresponding use of the explicit generators l_{ab} imposes a much more severe limitation on the representations of $R(4)$. By generalizing the l_{ab}, the condition $\mathbf{L}\cdot\mathbf{A} = 0$ can be relaxed, and the additional representations of $R(4)$ appear. An example of this occurs in the analysis of the dynamics of two interacting particles possessing both electric and magnetic charges [17, 18].

2.3 QUATERNIONS

The two commuting vectors $\boldsymbol{l}$ and $\boldsymbol{\lambda}$ of Section 1.6 are an obvious example of the generators of $R(4)$. If they are taken in the form of Eqs.

(1.22)–(1.23), however, it is not at all clear how the four-dimensional point $(x_1x_2x_3x_4)$ is related to the Euler angles $(\alpha\beta\gamma)$. To explore this connection, we introduce, in addition to the imaginary number i, the two extraordinaries j and k for which

$$\mathrm{i}^2 = \mathrm{j}^2 = \mathrm{k}^2 = -1, \quad \mathrm{ij} = -\mathrm{ji} = \mathrm{k}, \quad \mathrm{jk} = -\mathrm{kj} = \mathrm{i}, \quad \mathrm{ki} = -\mathrm{ik} = \mathrm{j}. \tag{2.7}$$

A quantity of the type

$$Q = q_4 + \mathrm{i}q_1 + \mathrm{j}q_2 + \mathrm{k}q_3$$

is a quaternion. It is sometimes convenient to separate the scalar and vector parts by writing [19]

$$S(Q) = q_4, \qquad V(Q) = \mathrm{i}q_1 + \mathrm{j}q_2 + \mathrm{k}q_3.$$

In these terms, the conjugate quaternion Q^* is defined as $S(Q) - V(Q)$.

The relevance of quaternions to rotating frames becomes apparent as soon as it is noticed that the transformation (1.10) is expressed by the equation [20]

$$\mathrm{i}\xi + \mathrm{j}\eta + \mathrm{k}\zeta = e^{-\mathrm{k}\gamma/2}e^{-\mathrm{j}\beta/2}e^{-\mathrm{k}\alpha/2}(\mathrm{i}x + \mathrm{j}y + \mathrm{k}z)e^{\mathrm{k}\alpha/2}e^{\mathrm{j}\beta/2}e^{\mathrm{k}\gamma/2}. \tag{2.8}$$

It will now be shown that the connection between a rotation characterized by ω and the point $(x_1x_2x_3x_4)$ is given by

$$e^{-\mathrm{k}\gamma/2}e^{-\mathrm{j}\beta/2}e^{-\mathrm{k}\alpha/2} = x_4 + \mathrm{i}x_1 + \mathrm{j}x_2 + \mathrm{k}x_3. \tag{2.9}$$

To do this, we write (2.8) in the quaternionic form

$$\rho = xrx^*, \tag{2.10}$$

and then evaluate $\boldsymbol{\lambda}$ by means of Eq. (1.21). The partial derivatives $\partial/\partial\xi$, $\partial/\partial\eta$, and $\partial/\partial\zeta$ that occur here carry the implication that the quaternion r is constant; so we have merely to relate these derivatives to those of the type $\partial/\partial x_a$ by means of Eq. (2.10). One minor difficulty arises: there are four variables x_a but only three components ξ, η, and ζ of ρ. This can be readily understood, since $(x_1x_2x_3x_4)$ lies on the unit sphere in four dimensions. From a formal point of view, however, we need to augment ρ by one variable. This is easily done by defining

$$\rho = \epsilon + \mathrm{i}\xi + \mathrm{j}\eta + \mathrm{k}\zeta.$$

Several other preliminary statements need to be made. By expanding the quaternions in detail, the following three equations can be established for quaternions P and Q:

$$S(Q) - \mathrm{i}S(\mathrm{i}Q) - \mathrm{j}S(\mathrm{j}Q) - \mathrm{k}S(\mathrm{k}Q) = Q, \tag{2.11}$$

$$S(PQ) + \mathrm{i}S(P\mathrm{i}Q) + \mathrm{j}S(P\mathrm{j}Q) + \mathrm{k}S(P\mathrm{k}Q) = P^*Q^*, \tag{2.12}$$

$$P^*Q^* = (QP)^*. \tag{2.13}$$

With the two definitions

$$\nabla_x = \frac{\partial}{\partial x_4} + \mathrm{i}\frac{\partial}{\partial x_1} + \mathrm{j}\frac{\partial}{\partial x_2} + \mathrm{k}\frac{\partial}{\partial x_3},$$

$$\nabla_\rho = \frac{\partial}{\partial \epsilon} + \mathrm{i}\frac{\partial}{\partial \xi} + \mathrm{j}\frac{\partial}{\partial \eta} + \mathrm{k}\frac{\partial}{\partial \zeta},$$

we can begin by noticing that

$$\frac{\partial}{\partial x_1} = -S\left(\frac{\partial \rho}{\partial x_1}\nabla_\rho\right) = -S(\{\mathrm{i}rx^* - xr\mathrm{i}\}\nabla_\rho),$$

from Eq. (2.10). Similar equations hold for the other derivatives. Thus, using Eqs. (2.11)–(2.13), we have

$$\begin{aligned}\nabla_x^* &= -S(\{rx^* + xr\}\nabla_\rho) + \mathrm{i}S(\{\mathrm{i}rx^* - xr\mathrm{i}\}\nabla_\rho)\\ &\quad + \mathrm{j}S(\{\mathrm{j}rx^* - xr\mathrm{j}\}\nabla_\rho) + \mathrm{k}S(\{\mathrm{k}rx^* - xr\mathrm{k}\}\nabla_\rho)\\ &= -rx^*\nabla_\rho - (xr)^*\nabla_\rho^*\\ &= -rx^*\nabla_\rho + rx^*\nabla_\rho^*,\end{aligned}$$

since $r^* = -r$. Hence

$$x\nabla_x^* = -\rho\nabla_\rho + \rho\nabla_\rho^* = -2\rho V(\nabla_\rho).$$

By expanding the quaternions, we quickly see that

$$V(x\nabla_x^*) = \mathrm{i}(iA_x - iL_x) + \mathrm{j}(iA_y - iL_y) + \mathrm{k}(iA_z - iL_z).$$

On setting $\epsilon = 0$, we find

$$V(\rho V(\nabla_\rho)) = \mathrm{i}(i\lambda_\xi) + \mathrm{j}(i\lambda_\eta) + \mathrm{k}(i\lambda_\zeta).$$

We can conclude that

$$\boldsymbol{\lambda} = -\tfrac{1}{2}(\mathbf{A} - \mathbf{L}) = \mathbf{K}_2.$$

A precisely similar analysis yields $\boldsymbol{l} = \mathbf{K}_1$. These results confirm that the rotation characterized by ω is represented by the point x on the four-dimensional sphere according to Eq. (2.9). By equating components, the following explicit expressions are obtained:

$$\begin{aligned}x_1 &= \sin\tfrac{1}{2}\beta\sin\tfrac{1}{2}(\alpha - \gamma),\\ x_2 &= -\sin\tfrac{1}{2}\beta\cos\tfrac{1}{2}(\alpha - \gamma),\\ x_3 &= -\cos\tfrac{1}{2}\beta\sin\tfrac{1}{2}(\alpha + \gamma),\\ x_4 &= \cos\tfrac{1}{2}\beta\cos\tfrac{1}{2}(\alpha + \gamma).\end{aligned} \qquad (2.14)$$

The equal ranks of the double tensor $\mathbf{D}^{(JJ)}$ now appear as two identical irreducible representations of $R(3)$, as required from the condition (2.6).

2.4 FOUR-DIMENSIONAL SPHERICAL HARMONICS

The coordinate system (2.14) is very similar to one used by Bateman [21]; the space is broken into two by assigning $\sin \frac{1}{2}\beta$ and $\cos \frac{1}{2}\beta$ to the upper and lower parts, (x_1x_2) and (x_3x_4); and then these two parts are themselves split into two. Alternatively, we can set up a triangular structure by extending the ordinary spherical-polar scheme:

$$\begin{aligned} x_1 &= r \sin \chi \sin \theta \cos \phi, \\ x_2 &= r \sin \chi \sin \theta \sin \phi, \\ x_3 &= r \sin \chi \cos \theta, \\ x_4 &= r \cos \chi. \end{aligned} \tag{2.15}$$

If we take θ and ϕ to lie within their traditional bounds $(0 \leq \phi \leq 2\pi,$ $0 \leq \theta \leq \pi)$, then the surface of the hypersphere is covered in just one way provided $0 \leq \chi \leq \pi$. The triad $(\chi\theta\phi)$ is abbreviated to Ω. We can now evaluate $\mathbf{L}$ and $\mathbf{A}$ in terms of these coordinates. It is at once found that $\mathbf{L} = \boldsymbol{l}$, where $\boldsymbol{l}$ is defined as in Eqs. (1.6). As for $\mathbf{A}$, we get [22]

$$A_\pm = ie^{\pm i\phi}\left(\sin \theta \frac{\partial}{\partial \chi} + \cot \chi \cos \theta \frac{\partial}{\partial \theta} \pm i \cot \chi \csc \theta \frac{\partial}{\partial \phi}\right),$$

$$A_z = i\left(\cos \theta \frac{\partial}{\partial \chi} - \cot \chi \sin \theta \frac{\partial}{\partial \theta}\right). \tag{2.16}$$

In three dimensions, the operators $\boldsymbol{l}^2$ and l_z form a commuting pair, and the corresponding quantum numbers (l and m) are available for labeling the spherical harmonics $Y_{lm}(\theta, \phi)$. It is straightforward to show that $\mathbf{k}^2$ (of Eq. 2.6) commutes with $\boldsymbol{l}^2$ and l_z, so the three quantum numbers (klm) can be used in four dimensions. The generators $\boldsymbol{l}$ and $\mathbf{A}$ of $R(4)$ commute with $\mathbf{k}^2$, which implies that irreducible representations of $R(4)$ correspond to given k values but various l and m. Since the two commuting angular momenta $\mathbf{K}_1$ and $\mathbf{K}_2$ satisfy $\mathbf{K}_1 + \mathbf{K}_2 = \mathbf{L}\ (= \boldsymbol{l})$, the allowed l values are $0, 1, 2, \ldots, 2k$. This sequence corresponds to the level of the nonrelativistic hydrogen atom for which $n = 2k + 1$. In anticipation of future applications, we write the four-dimensional spherical harmonics as $Y_{nlm}(\Omega)$, it being understood that $k = \frac{1}{2}(n - 1)$.

The three-dimensional harmonics $Y_{lm}(\theta, \phi)$ appear when Laplace's equation $\nabla^2 \psi = 0$ is examined for solutions of the form $r^l f(\theta, \phi)$. The N-dimensional problem has been analyzed by Hill [23] and again by Louck [24]. The generalized Laplacian can be written

$$\sum_a \frac{\partial^2}{\partial x_a{}^2} = \frac{1}{r^{N-1}} \frac{\partial}{\partial r}\left(r^{N-1} \frac{\partial}{\partial r}\right) - \frac{1}{r^2} \sum_{b>a} l_{ab}{}^2 \tag{2.17}$$

(see Problem 2.1). If we seek a solution of the form $r^j Y(\Omega)$, it is found that $Y(\Omega)$ must satisfy

$$\sum_{b>a} l_{ab}{}^2 Y(\Omega) = j(j + N - 2) Y(\Omega). \tag{2.18}$$

A function satisfying this equation is a hyperspherical harmonic [25]. In the case of $N = 4$, for which $Y = Y_{nlm}$, we have

$$\sum_{b>a} l_{ab}{}^2 = \boldsymbol{l}^2 + \mathbf{A}^2 = 4\mathbf{k}^2,$$

so $Y_{nlm}(\Omega)$ is an eigenfunction of $\mathbf{k}^2$ with eigenvalue $\frac{1}{4}j(j + 2)$. Thus $j = 2k = n - 1$. To ensure that $Y_{nlm}(\Omega)$ has the correct properties with respect to $\boldsymbol{l}$, we factor out $Y_{lm}(\theta, \phi)$. The remaining part can be found by referring to Eq. (2.18) and following Hill's analysis. To describe the results, we insert a phase factor $(-i)^l$ and write

$$Y_{nlm}(\Omega) = (-i)^l C_{n,l}(\chi) Y_{lm}(\theta, \phi), \tag{2.19}$$

where

$$C_{n,l}(\chi) = \left(\frac{2n(n - l - 1)!}{\pi(n + l)!}\right)^{1/2} \sin^l \chi \left(\frac{d}{d \cos \chi}\right)^l C_{n-1}(\cos \chi), \tag{2.20}$$

in which the Gegenbauer polynomials C_t are defined by

$$\frac{1}{1 - 2\mu h + h^2} \equiv \sum_{t=0}^{\infty} h^t C_t(\mu). \tag{2.21}$$

Another form for the Gegenbauer polynomials is in common use:

$$\frac{1}{(1 - 2\mu h + h^2)^\nu} \equiv \sum_{t=0}^{\infty} h^t C_t{}^\nu(\mu). \tag{2.22}$$

In terms of these polynomials,

$$C_t(\mu) = C_t{}^1(\mu), \qquad P_t(\mu) = C_t{}^{1/2}(\mu).$$

It is straightforward to show that

$$\sum_{a=1}^{4} dx_a{}^2 = dr^2 + r^2 \sin^2 \chi \, d\theta^2 + r^2 \sin^2 \chi \sin^2 \theta \, d\phi^2 + r^2 \, d\chi^2,$$

so the infinitesimal area of hypersurface on the unit sphere is given by

$$d\Omega = \sin^2 \chi \sin \theta \, d\chi \, d\theta \, d\phi. \tag{2.23}$$

The Y_{nlm} are orthonormal in that

$$\int Y_{n'l'm'}^*(\Omega) Y_{nlm}(\Omega) \, d\Omega = \delta(l, l')\delta(m, m')\delta(n, n').$$

Equation (1.56) leads to

$$Y_{nlm}^*(\Omega) = (-1)^{l-m} Y_{nl-m}(\Omega). \tag{2.24}$$

The spherical harmonics $Y_{nlm}(\Omega)$ are identical to those introduced by Biedenharn [26], at least to within a phase independent of l and m. A brief tabulation is presented in Appendix 2.

2.5 PHASES

Two alternative descriptions are now available for a four-dimensional spherical harmonic. We can write it explicitly as $Y_{nlm}(\Omega)$ through Eq. (2.19); or we can simply describe the coupling by means of the ket $|(kk)lm\rangle$. It would be advantageous to work solely with the kets, since we could apply the techniques of angular momentum theory instead of having to use the properties of special functions, such as the Gegenbauer polynomials. We already know that the correspondence must be of the form

$$Y_{nlm}(\Omega) \equiv e^{i\sigma} |(kk)lm\rangle \qquad (n = 2k + 1);$$

the only uncertainty is in the phase angle σ. This can be found by matching the effects of the generators of $R(4)$ on both sides of this equivalence. We know that $Y_{lm}(\theta, \phi)$ and $|lm\rangle$ are correctly matched with respect to $\boldsymbol{l}$ from the analysis of Section 1.2; so we can find σ by considering components of the other generator, $\mathbf{A}$. Moreover, only one component, say A_z, need be considered, since the others can be derived from commutation with the previously tested operator $\boldsymbol{l}$.

Consider, then, the effect of A_z, as given in Eqs. (2.16), on $Y_{nlm}(\Omega)$. As might be anticipated after the remarks above, the analysis is tedious, and it seems best to relegate it to a problem (Problem 2.3). The result is

$$A_z Y_{nlm}(\Omega) = \left(\frac{[(l+1)^2 - m^2][n^2 - (l+1)^2]}{(2l+1)(2l+3)} \right)^{1/2} Y_{n,l+1,m}(\Omega)$$
$$+ \left(\frac{[l^2 - m^2][n^2 - l^2]}{(2l-1)(2l+1)} \right)^{1/2} Y_{n,l-1,m}(\Omega). \tag{2.25}$$

To illustrate the tensorial approach, we calculate the coefficient of $Y_{n,l+1,m}(\Omega)$, namely,

$$\langle (kk)l+1, m \mid A_z \mid (kk)lm \rangle.$$

We first note that

$$A_z = K_{1z} - K_{2z} = 2K_{1z} - l_z,$$

and since l_z is diagonal with respect to l, the last term can be dropped. Next, we observe that K_{1z} acts on the first part of the coupled system, so we use Eq. (1.66) after first applying the WE theorem:

$$\begin{aligned}
&\langle (kk)l+1, m \mid A_z \mid (kk)lm \rangle \\
&\quad = 2(-1)^{l+1-m} \begin{pmatrix} l+1 & 1 & l \\ -m & 0 & m \end{pmatrix} ((kk)l+1 \,\|\, K_1 \,\|\, (kk)l) \\
&\quad = 2(-1)^{2k-m} \{(2l+1)(2l+3)\}^{1/2} \\
&\quad\quad \times \begin{pmatrix} l+1 & 1 & l \\ -m & 0 & m \end{pmatrix} \begin{Bmatrix} k & 1 & k \\ l & k & l+1 \end{Bmatrix} (k \,\|\, K_1 \,\|\, k).
\end{aligned}$$

To complete the calculation, we put

$$(k \,\|\, K_1 \,\|\, k) = \{k(k+1)(2k+1)\}^{1/2},$$

which follows from Eq. (1.53), and turn to Appendix I for explicit expressions for the 3-j and 6-j symbols. The result is

$$\begin{aligned}
&\langle (kk)l+1, m \mid A_z \mid (kk)lm \rangle \\
&\quad = \left(\frac{[(l+1)^2 - m^2][(2k+1)^2 - (l+1)^2]}{(2l+1)(2l+3)} \right)^{1/2}.
\end{aligned}$$

On setting $2k+1 = n$, precisely the same result as before is obtained, not only as regards magnitude (which must necessarily be the case) but also for phase as well. The same turns out to be true for the second coefficient in Eq. (2.25). We may therefore set $\sigma = 0$:

$$Y_{nlm}(\Omega) \equiv \mid (kk)lm \rangle. \tag{2.26}$$

It was to obtain a result of this simplicity that the phase $(-i)^l$ was inserted in the definition of $Y_{nlm}(\Omega)$ given in Eq. (2.19).

Before leaving Eq. (2.25), it is impossible not to comment on the remarkable symmetry between n and m. In fact, if we write $n \to m$ and $m \to n$, the coefficients remain invariant. This is quite unexpected and particularly intriguing because n and m play such different roles in the analysis. The spectroscopists who defined the magnetic quantum number m and the principal quantum number n for the hydrogen atom were certainly not aware of any reciprocity between them. This kind of result gives us a glimpse of the extraordinarily rich group structure of the hydrogen atom, a subject that has been extensively studied, particularly in the last decade. The work of Barut and his collaborators [27] and of Armstrong [28] is especially relevant for us. We return to the symmetry of Eq. (2.25) in Section 4.12.

2.6 CG COEFFICIENTS FOR $R(4)$

We are now in a position to derive Biedenharn's formula [26] for the CG coefficients of $R(4)$. In analogy with $R(3)$, a CG coefficient for $R(4)$ is written as

$$(nlm,\ n'l'm' \mid n''l''m'').$$

If the $R(3) \times R(3)$ structure of $R(4)$ is made explicit, this becomes

$$((kk)lm,\ (k'k')l'm' \mid (k''k'')l''m''),$$

where $n = 2k + 1$, etc. If we now use the expansion

$$\langle (kk)lm,\ (k'k')l'm' \mid = \sum_{t,p} (lm,\ l'm' \mid tp) \langle (kk)l,\ (k'k')l',\ tp \mid,$$

we get

$$\begin{aligned} &((kk)lm,\ (k'k')l'm' \mid (k''k'')l''m'') \\ &\quad = (lm,\ l'm' \mid l''m'')((kk)l,\ (k'k')l',\ l'' \mid (kk')k'',\ (kk')k'',\ l'') \\ &\quad = (lm,\ l'm' \mid l''m'')(2k''+1)[l,\ l']^{1/2} \begin{Bmatrix} k & k & l \\ k' & k' & l' \\ k'' & k'' & l'' \end{Bmatrix}. \end{aligned}$$

So we conclude that

$$\begin{aligned} &(nlm,\ n'l'm' \mid n''l''m'') \\ &\quad = (lm,\ l'm' \mid l''m'')n''[l,\ l']^{1/2} \begin{Bmatrix} \frac{1}{2}(n-1) & \frac{1}{2}(n-1) & l \\ \frac{1}{2}(n'-1) & \frac{1}{2}(n'-1) & l' \\ \frac{1}{2}(n''-1) & \frac{1}{2}(n''-1) & l'' \end{Bmatrix}. \end{aligned} \tag{2.27}$$

2.7 COMBINATION THEOREM FOR GEGENBAUER POLYNOMIALS

A common problem in many branches of quantum mechanics is the resolution of a product of two special functions into a sum of special functions of the same kind. The situation when all functions share a common argument is of particular interest. Usually several possible approaches are available to us. The obvious method is simply to take advantage of the orthonormality properties of the special functions. This is illustrated in Problem 2.4 for the Gegenbauer polynomials C_t. However, the special nature of the C_t polynomials makes it possible to give an alternative and very much simpler derivation.

The object of the analysis is to find the numbers $\Delta(tt's)$ in the expansion

$$C_t(\mu)C_{t'}(\mu) = \sum_s \Delta(tt's)C_s(\mu),$$

where $\mu = \cos\chi$. From Eq. (2.21), we have

$$\begin{aligned}\sum_t h^t C_t(\cos\chi) &= (1 - he^{i\chi} - he^{-i\chi} + h^2)^{-1}\\ &= \{(1 - he^{i\chi})(1 - he^{-i\chi})\}^{-1}\\ &= (1 + he^{i\chi} + h^2e^{2i\chi} + \cdots)(1 + he^{-i\chi} + h^2e^{-2i\chi} + \cdots).\end{aligned}$$

Thus

$$C_t(\cos\chi) = e^{it\chi} + e^{i(t-2)\chi} + \cdots + e^{-it\chi}. \tag{2.28}$$

We can now construct the product $C_tC_{t'}$ by direct multiplication and then proceed to decompose it into the parts C_s. However, we can go immediately to the result simply by observing that the right-hand side of Eq. (2.28) is the character of an irreducible representation of $R(3)$ corresponding to an angle of rotation 2χ and an angular momentum quantum number $\frac{1}{2}t$ (see, for example, Wigner [10]). The two angular momenta $\frac{1}{2}t$ and $\frac{1}{2}t'$ can couple to produce the resultants

$$\tfrac{1}{2}(t + t'),\ \tfrac{1}{2}(t + t') - 1,\ \tfrac{1}{2}(t + t') - 2,\ \ldots,\ \tfrac{1}{2}\,|\,t - t'\,|,$$

each of which occurs just once. Thus

$$\Delta(tt's) = 1$$

if t, t', and s satisfy the triangular condition and if the sum $t + t' + s$ is even: otherwise $\Delta(tt's) = 0$. The origin of the angular momentum $\frac{1}{2}t$ lies in the collapse of Y_{nlm} to C_{n-1} when $l = m = 0$; for then $n - 1 = 2k = t$, and so $\frac{1}{2}t = k$.

It might be worthwhile to illustrate this result with an example. We find, from the generating function,

$$C_0(\mu) = 1, \qquad C_3(\mu) = 8\mu^3 - 4\mu,$$
$$C_1(\mu) = 2\mu, \qquad C_4(\mu) = 16\mu^4 - 12\mu^2 + 1,$$
$$C_2(\mu) = 4\mu^2 - 1, \qquad C_5(\mu) = 32\mu^5 - 32\mu^3 + 6\mu,$$

and

$$\begin{aligned} C_2C_3 &= (4\mu^2 - 1)(8\mu^3 - 4\mu) \\ &= 32\mu^5 - 24\mu^3 + 4\mu = C_1 + C_3 + C_5. \end{aligned}$$

2.8 COMBINATION THEOREMS FOR SPHERICAL HARMONICS

The analogous combination theorem for the spherical harmonics $Y_{lm}(\theta, \phi)$ can be found equally easily. We set $a' = b' = 0$ in Eq. (1.41), thereby getting the equation

$$\mathfrak{D}_{m0}{}^{l}(\omega)\mathfrak{D}_{m'0}{}^{l'}(\omega) = \sum_{l''m''} (lm, l'm' \mid l''m'')(l0, l'0 \mid l''0)\mathfrak{D}_{m''0}{}^{l''}(\omega).$$

From Problem 1.4, we have

$$\mathfrak{D}_{m0}{}^{l}(-\phi\theta\gamma) = C_m{}^{(l)}(\theta, -\phi)^* = C_m{}^{(l)}(\theta\phi) = \{4\pi/(2l + 1)\}^{1/2}Y_{lm}(\theta, \phi),$$

and so

$$\begin{aligned} Y_{lm}(\theta, \phi)Y_{l'm'}(\theta, \phi) = \sum_{l''m''} &\{(2l + 1)(2l' + 1)/4\pi(2l'' + 1)\}^{1/2} \\ &\times (l0, l'0 \mid l''0)(lm, l'm' \mid l''m'')Y_{l''m''}(\theta, \phi). \end{aligned} \tag{2.29}$$

This equation assumes a much more concise form with a little manipulation. First, the spherical harmonics are converted to **C** tensors; both sides of the equation are then multiplied by $(lm, l'm' \mid LM)$ and sums over m and m' are performed. The result is

$$(\mathbf{C}^{(l)}\mathbf{C}^{(l')})^{(L)} = (-1)^L(2L + 1)^{1/2}\begin{pmatrix} l & l' & L \\ 0 & 0 & 0 \end{pmatrix}\mathbf{C}^{(L)}. \tag{2.30}$$

A similar procedure is available for $R(4)$. As Talman [29] shows, analogs of the $\mathfrak{D}$ functions can be found, and they too can be made to reduce to $Y_{nlm}(\Omega)$. It seems preferable, however, to take the opportunity of illustrating a different method. As a preliminary to this, we note that the dependence on m, m', and m'' of the coefficient in Eq. (2.29) is simply a CG coefficient. This must be so, since the m quantum numbers are the component labels of tensors, and the correct coupling must be exhibited. In $R(4)$, both l and m play the role of components, so that the parallel equation to Eq. (2.29) takes the form

$$Y_{nlm}(\Omega)\,Y_{n'l'm'}(\Omega) = \sum_{n''l''m''} f(nn'n'')\,(nlm,\, n'l'm' \mid n''l''m'')\,Y_{n''l''m''}(\Omega). \tag{2.31}$$

The function $f(nn'n'')$ can be determined by selecting a suitable set of values for the components. If we choose $l = l' = m = m' = 0$, then the sum reduces to a sum over n'' only, since $l'' = m'' = 0$. It is not difficult to show that

$$Y_{n00}(\Omega) = (2\pi^2)^{-1/2}C_{n-1}(\cos\chi).$$

Furthermore,

$$(n00,\, n'00 \mid n''00) = (00,\, 00 \mid 00)\,n'' \begin{Bmatrix} \frac{1}{2}(n-1) & \frac{1}{2}(n-1) & 0 \\ \frac{1}{2}(n'-1) & \frac{1}{2}(n'-1) & 0 \\ \frac{1}{2}(n''-1) & \frac{1}{2}(n''-1) & 0 \end{Bmatrix}$$

$$= (n''/nn')^{1/2}\Delta(n-1,\, n'-1,\, n''-1),$$

where $\Delta = 1$ if the triad $(n-1,\, n'-1,\, n''-1)$ satisfies the triangular condition and if, in addition, $n + n' + n''$ is odd: otherwise $\Delta = 0$. Thus Δ is precisely the same as the function Δ introduced in Section 2.7. So Eq. (2.31) becomes

$$C_{n-1}(\cos\chi)\,C_{n'-1}(\cos\chi) = \sum_{n''} (2\pi^2 n''/nn')^{1/2} f(nn'n'')$$
$$\times\, \Delta(n-1,\, n'-1,\, n''-1)\,C_{n''-1}(\cos\chi).$$

For this to be identical to the expansion obtained in Section 2.7, we must have

$$f(nn'n'') = (nn'/2\pi^2 n'')^{1/2}.$$

This can now be fed back into Eq. (2.31). At the same time, the explicit form 2.27 can be used for the CG coefficient. The result is

$$Y_{nlm}(\Omega)\,Y_{n'l'm'}(\Omega)$$
$$= \sum_{n''l''m''} \{(2l+1)(2l'+1)nn'n''/2\pi^2\}^{1/2}(lm,\, l'm' \mid l''m'')$$
$$\times \begin{Bmatrix} \frac{1}{2}(n-1) & \frac{1}{2}(n-1) & l \\ \frac{1}{2}(n'-1) & \frac{1}{2}(n'-1) & l' \\ \frac{1}{2}(n''-1) & \frac{1}{2}(n''-1) & l'' \end{Bmatrix} Y_{n''l''m''}(\Omega). \tag{2.32}$$

The 9-j symbol in this equation automatically vanishes when $\Delta(n-1, n'-1, n''-1) = 0$, so our derivation of $f(nn'n'')$ holds good in the general case.

From a tensorial point of view, Y_{nlm} is a component of $\mathbf{Y}^{(kk)l}$, where the three ranks exhibit the properties of the spherical harmonic with respect to $\mathbf{K}_1$, $\mathbf{K}_2$, and their resultant, $\boldsymbol{l}$. Using this notation, Eq. (2.32) can be more concisely expressed by

$$\{\mathbf{Y}^{(kk)l}(\Omega)\mathbf{Y}^{(k'k')l'}(\Omega)\}^{(l'')}$$
$$= \sum_{k''l''} \sqrt{\frac{nn'n''(2l+1)(2l'+1)}{2\pi^2}} \begin{Bmatrix} k & k & l \\ k' & k' & l' \\ k'' & k'' & l'' \end{Bmatrix} \mathbf{Y}^{(k''k'')l''}(\Omega). \tag{2.33}$$

2.9 ADDITION THEOREMS

In the previous two sections, the functions being combined depend on the same angular variables. This makes it reasonably straightforward to apply tensor analysis. Another class of theorems, usually called *addition* theorems, describe how functions of dissimilar arguments can be combined. The origin of these theorems is a composition rule for group elements. For example, in Eq. (1.47), two successive rotations, $D(\omega_2)$ and $D(\omega_1)$, characterized by Euler triads ω_2 and ω_1, are expressed as a single rotation $D(\omega)$. Symbolically, we can represent the construction of ω from ω_1 and ω_2 by writing $\boldsymbol{\omega} = \boldsymbol{\omega}_1 + \boldsymbol{\omega}_2$. If we wish to have $\boldsymbol{\omega}$ represent the difference between two rotations, i.e., to have $\boldsymbol{\omega} = \boldsymbol{\omega}_1 - \boldsymbol{\omega}_2$, then

$$D(\omega) = D(\omega_1)D(\omega_2)^{-1}.$$

This is often a more convenient starting point. Taking matrix elements between $\langle kp \mid$ and $\mid kq\rangle$ gives

$$\begin{aligned} \mathfrak{D}_{pq}{}^k(\alpha\beta\gamma) &= \sum_t \mathfrak{D}_{pt}{}^k(\alpha_1\beta_1\gamma_1)\mathfrak{D}_{tq}{}^k(-\gamma_2\ -\beta_2\ -\alpha_2) \\ &= \sum_t \mathfrak{D}_{pt}{}^k(\alpha_1\beta_1\gamma_1)\mathfrak{D}_{-q-t}{}^k(\alpha_2\beta_2\gamma_2)(-1)^{t-q}. \end{aligned} \tag{2.34}$$

Use has been made of Eqs. (1.12)–(1.13) in deriving this result.

Several addition theorems can be obtained as special cases of Eq. (2.34). The most familiar of these follows by putting $p = q = 0$ and using the results of Problem 1.4. To throw the result into the usual form, we make the substitutions

$$\beta = \theta, \quad \beta_1 = \theta_1, \quad \beta_2 = \theta_2, \quad \gamma = -\phi, \quad \gamma_1 = -\phi_1, \quad \gamma_2 = -\phi_2,$$
$$k = l, \quad t = -m,$$

and use Eq. (1.56) in the form

$$Y_{l-m}(\theta_1, \phi_1) = (-1)^m Y_{lm}^*(\theta_1, \phi_1).$$

We get

$$P_l(\cos\theta) = 4\pi(2l+1)^{-1} \sum_m Y_{lm}^*(\theta_1, \phi_1) Y_{lm}(\theta_2, \phi_2). \tag{2.35}$$

We have now to interpret θ in terms of θ_1, ϕ_1, θ_2, and ϕ_2. Suppose $D(\omega_1)$ takes a unit vector $\mathbf{r}_0$ into $\mathbf{r}_1$; that is, $\mathbf{r}_0 \to \mathbf{r}_1$. Then $D(\omega_2)$ corresponds to $\mathbf{r}_0 \to \mathbf{r}_2$, and $D(\omega_1)D(\omega_2)^{-1}$ corresponds to $\mathbf{r}_2 \to \mathbf{r}_1$. This transformation is effected by $D(\omega)$, and thus depends on θ and ϕ. However, ϕ does not appear on the left-hand side of Eq. (2.35), which implies that the entire rotation can be described in terms of θ alone. From elementary geometry,

$$\cos\theta = \mathbf{r}_1 \cdot \mathbf{r}_2 = \cos\theta_1 \cos\theta_2 + \sin\theta_1 \sin\theta_2 \cos(\phi_1 - \phi_2).$$

In tensorial notation, Eq. (2.35) takes the simple form

$$P_l(\cos\theta) = (\mathbf{C}_1^{(l)} \cdot \mathbf{C}_2^{(l)}). \tag{2.36}$$

Addition theorems of a more complicated kind can also be derived from Eq. (2.34). If we pass to the double tensors $\mathbf{D}^{(kk)}$, defined by Eq. (1.27), then Eq. (2.34) becomes

$$D_{p-q}{}^{(kk)}(\omega) = \sum_t (-1)^{k+t}(2k+1)^{-1/2} D_{p-t}{}^{(kk)}(\omega_1) D_{-q\,t}{}^{(kk)}(\omega_2).$$

Both sides of this equation are now multiplied by $(kp, k\ -q \mid lm)$ and summed over p and q. Equation (1.34) is used to perform the sum over t, with the result

$$\mathbf{D}^{(kk)l}(\omega) = \{\mathbf{D}^{(kk)}(\omega_1)\mathbf{D}^{(kk)}(\omega_2)\}^{(l0)l}.$$

Of course, this is nothing more than Eq. (2.34) in tensorial form. The coupling on the right can be rearranged by introducing the recoupling coefficient

$$((kk)l, (kk)0, l \mid (kk)l_1, (kk)l_2, l)$$
$$= [l, l_1, l_2]^{1/2} \begin{Bmatrix} k & k & l \\ k & k & 0 \\ l_1 & l_2 & l \end{Bmatrix} = (-1)^{2k+l_2+l} \sqrt{\frac{[l_1, l_2]}{[k]}} \begin{Bmatrix} l_1 & l_2 & l \\ k & k & k \end{Bmatrix},$$

whence we get

$$\mathbf{D}^{(kk)l}(\omega) = \sum_{l_1,l_2} (-1)^{2k+l_2+l} \sqrt{\frac{[l_1, l_2]}{[k]}} \begin{Bmatrix} l_1 & l_2 & l \\ k & k & k \end{Bmatrix} \times \{\mathbf{D}^{(kk)l_1}(\omega_1)\mathbf{D}^{(kk)l_2}(\omega_2)\}^{(l)}. \tag{2.37}$$

This result can be used to get an addition theorem for four-dimensional spherical harmonics. By comparing Eqs. (2.14) and (2.15), we can, in principle, relate the Euler angles ω to the spherical polars Ω through the intermediary of the point $(x_1x_2x_3x_4)$. This is unnecessary to carry out in detail, since the coupling exhibited by $D_m^{(kk)l}$ is identical to that implied by Y_{nlm} if we take $n = 2k + 1$. As far as the generators of $R(4)$ are concerned, a direct relation must exist between these two functions. We know from Section 2.5 that the phase of the spherical harmonic follows that of the coupling scheme $|(kk)lm\rangle$, so the relative phase between D and Y can at most depend on k. The Y functions are normalized, whereas $D_m^{(kk)l}$, like $D_{pq}^{(kk)}$, requires the prefixed coefficient $(8\pi^2)^{-1/2}$. We must also note that the infinitesimal areas on the hypersphere are of different magnitude, since

$$\int d\omega = 8\pi^2, \qquad \int d\Omega = 2\pi^2.$$

Thus the relation must run

$$D_m^{(kk)l}(\omega) = e^{i\tau}(2\pi^2)^{1/2}Y_{nlm}(\Omega), \tag{2.38}$$

where $n = 2k + 1$ and where τ depends on n (or k) only. We have only to compare the two mutually consistent schemes

$$\alpha = \beta = \gamma = 0, \qquad \chi = \theta = \phi = 0,$$

to find that we can set $\tau = 0$. Furthermore, the relation $\boldsymbol{\omega} = \boldsymbol{\omega}_1 - \boldsymbol{\omega}_2$ implies $\boldsymbol{\Omega} = \boldsymbol{\Omega}_1 - \boldsymbol{\Omega}_2$. It only remains to substitute for the $\mathbf{D}$ tensors in Eq. (2.37), and we arrive at an addition theorem for four-dimensional spherical harmonics:

$$Y_{nlm}(\Omega) = \sum_{l_1l_2m_1m_2} (-1)^{n+l_2+l+1} \sqrt{\frac{2\pi^2\,[l_1, l_2]}{n}} \times \begin{Bmatrix} l_1 & l_2 & l \\ \frac{1}{2}(n-1) & \frac{1}{2}(n-1) & \frac{1}{2}(n-1) \end{Bmatrix} \times (l_1m_1, l_2m_2 \mid lm)\, Y_{nl_1m_1}(\Omega_1)\, Y_{nl_2m_2}(\Omega_2). \tag{2.39}$$

An alternative derivation, based on the four-dimensional rotation matrices, has been sketched by Alper [30]. It is worth noting, in passing, that Stone [25] has used an equation similar to (2.38) to derive explicit forms for certain CG coefficients for $R(3)$.

An important specialization of Eq. (2.39) occurs when $l = 0$. In this case,

$$Y_{n00}(\Omega) = (2\pi^2)^{-1/2} C_{n-1}(\cos \chi).$$

The notation is simplified by making the replacements

$$m_2 \to m, \quad m_1 \to -m, \quad l_2 \to l, \quad l_1 \to l.$$

With the aid of Eq. (2.24) we get

$$C_{n-1}(\cos \chi) = (2\pi^2/n) \sum_{lm} Y_{nlm}^*(\Omega_1) Y_{nlm}(\Omega_2). \tag{2.40}$$

This is the four-dimensional analog of Eq. (2.35). Corresponding to $\cos \theta$, we now have

$$\cos \chi = \sum_{a=1}^{4} x_{1a} x_{2a}, \tag{2.41}$$

where x_{1a} and x_{2a} are related to the angles $(\chi_1\theta_1\phi_1)$ and $(\chi_2\theta_2\phi_2)$ through equations of the type (2.15).

2.10 EXPANSIONS ON AND OFF THE HYPERSPHERE

There is no compulsion, when working in four-dimensional space, to limit attention to points on the surface of the unit hypersphere any more than there is, in ordinary three-dimensional space, to be restricted to points for which $|\mathbf{r}| = 1$. In three dimensions, an important quantity is r_{12}^{-1}, since it is proportional to the potential energy of two charges at the points $\mathbf{r}_1$ and $\mathbf{r}_2$. To expand r_{12}^{-1}, we introduce $r_<$ and $r_>$, the lesser and greater respectively of r_1 and r_2. Then

$$\begin{aligned} r_{12}^{-1} &= (r_>^2 + r_<^2 - 2r_>r_< \cos \theta)^{-1/2} = \sum_{k\geq 0} (r_<^k/r_>^{k+1}) P_k(\cos \theta) \\ &= \sum_{k\geq 0} (r_<^k/r_>^{k+1}) (\mathbf{C}_1^{(k)} \cdot \mathbf{C}_2^{(k)}) \end{aligned} \tag{2.42}$$

from Eq. (2.36). To exploit the analogy between Eqs. (2.35) and (2.40), we need to take r_{12}^{-2} when $\mathbf{r}_1$ and $\mathbf{r}_2$ are vectors in four-dimensional space:

$$\begin{aligned} r_{12}^{-2} &= (r_>^2 + r_<^2 - 2r_>r_< \cos \chi)^{-1} = \sum_{n\geq 1} (r_<^{n-1}/r_>^{n+1}) C_{n-1}(\cos \chi) \\ &= \sum_{n\geq 1} (2\pi^2/n)(r_<^{n-1}/r_>^{n+1}) \sum_{l,m} Y_{nlm}^*(\Omega_1) Y_{nlm}(\Omega_2). \end{aligned} \tag{2.43}$$

The angles Ω_1 and Ω_2 specify the directions of $\mathbf{r}_1$ and $\mathbf{r}_2$. If $|\mathbf{r}_1| = |\mathbf{r}_2| = 1$, we obtain

$$r_{12}^{-2} = \tfrac{1}{4}\csc^2 \tfrac{1}{2}\chi = \sum_{nlm} (2\pi^2/n)\, Y_{nlm}^*(\Omega_1)\, Y_{nlm}(\Omega_2). \tag{2.44}$$

PROBLEMS

2.1 The N coordinates x_a $(1 \le a \le N)$ for an N-dimensional Euclidean space are defined in terms of a radial coordinate r and $N - 1$ angular coordinates Ω through the equations $x_a = rf_a(\Omega)$. Prove that

$$\frac{\partial}{\partial r} = \sum_a \frac{x_a}{r}\frac{\partial}{\partial x_a}.$$

Extend the calculation to obtain the result

$$\frac{1}{r^{N-1}}\frac{\partial}{\partial r}\left(r^{N-1}\frac{\partial}{\partial r}\right) = \frac{N-1}{r^2}\sum_a x_a \frac{\partial}{\partial x_a} + \frac{1}{r^2}\sum_{ab} x_a x_b \frac{\partial^2}{\partial x_a\, \partial x_b},$$

and hence derive Eq. (2.17).

2.2 Prove

$$(1 - \mu^2)^{(1/2)l} C_{n-l-1}^{l+1}(\mu)\, 2^l l! [2n(n - l - 1)!/\pi(n + l)!]^{1/2} = C_{n,l}(\mu).$$

2.3 Obtain the following equations:

(1) $$\sin\theta \frac{\partial}{\partial\theta} Y_{lm}(\theta, \phi) = -(1 - \mu^2)\frac{d}{d\mu} Y_{lm}(\theta, \phi) \qquad (\mu \equiv \cos\theta)$$

$$= l\sqrt{\frac{(l+1)^2 - m^2}{(2l+1)(2l+3)}}\, Y_{l+1,m}(\theta, \phi)$$

$$- (l+1)\sqrt{\frac{l^2 - m^2}{(2l-1)(2l+1)}}\, Y_{l-1,m}(\theta, \phi),$$

(2) $$\cos\theta Y_{lm}(\theta, \phi)$$

$$= \sqrt{\frac{(l+1)^2 - m^2}{(2l+1)(2l+3)}}\, Y_{l+1,m}(\theta, \phi)$$

$$+ \sqrt{\frac{l^2 - m^2}{(2l-1)(2l+1)}}\, Y_{l-1,m}(\theta, \phi),$$

(3) $$\frac{d}{d\chi}\left[\sin^l \chi \left(\frac{d}{d\cos\chi}\right)^l C_{n-1}(\cos\chi)\right]$$

$$= \cot\chi\left[l \sin^l \chi \left(\frac{d}{d\cos\chi}\right)^l C_{n-1}(\cos\chi)\right]$$

$$- \sin^{l+1}\chi \left(\frac{d}{d\cos\chi}\right)^{l+1} C_{n-1}(\cos\chi),$$

(4) $$(1-\mu^2)\frac{d^{l+1}}{d\mu^{l+1}} C_{n-1}(\mu) - (2l+1)\mu \frac{d^l}{d\mu^l} C_{n-1}(\mu)$$

$$= -(n^2 - l^2)\frac{d^{l-1}}{d\mu^{l-1}} C_{n-1}(\mu).$$

Assemble these results to prove Eq. (2.25).

2.4 By multiplying both sides of the equation

$$C_t(\mu)C_{t'}(\mu) = \sum_s \Delta(tt's)C_s(\mu) \qquad (\mu \equiv \cos\chi)$$

by $C_{t''}(\cos\chi)\sin^2\chi\, d\chi$ and integrating over χ, prove that the coefficient of $h^t g^{t'} f^s$ in

$$\int_{-1}^{1} (1-\mu^2)^{1/2}(1-2\mu h + h^2)^{-1}(1-2\mu g+g^2)^{-1}(1-2\mu f+f^2)^{-1}\, d\mu$$

is $\frac{1}{2}\pi\Delta(tt's)$. Show that this is equal to the coefficient of $h^t g^{t'} f^s$ in

$$\tfrac{1}{2}\pi\{(1-gh)(1-hf)(1-fg)\}^{-1}$$

and deduce that $\Delta(tt's) = 1$ if t, t', and s satisfy the triangular condition and if, in addition, $t + t' + s$ is even: otherwise $\Delta(tt's) = 0$.

2.5 Derive Eq. (2.40) from the requirement that $C_{n-1}(\cos\chi)$, being a function of an angle between two radii vectors $\mathbf{r}_1$ and $\mathbf{r}_2$ of the hypersphere, is a scalar under rotations generated by $(J_{ab})_1 + (J_{ab})_2$.

3

$R(4)$ *in Physical Systems*

3.1 THE RIGID ROTATOR

The notion of a rigid rotator is an abstraction. Its significance for us lies in the fact that, when treating the motion of a physical object in quantum mechanics, we usually separate out the coordinates ω that refer specifically to the orientation of the object from those that represent internal motion. The resulting differential equation satisfied by the Euler angles ω is identical to that occurring in the quantum treatment of a classical rigid rotator. For this reason, the properties of this rather artificial object are relevant to us. Casimir [20] and van Winter [31] have given more complete treatments than the sketch presented here.

The rigid rotator is visualized as an object possessing three principal moments of inertia, I_1, I_2, and I_3. It is convenient to fix a frame F' in the rotator so that the ξ, η, and ζ axes coincide with the principal axes. A point $(\xi\eta\zeta)$ embedded in the rotator can be assigned coordinates (xyz) in a laboratory-fixed frame F. The connection with the Euler angles ω $(= \alpha\beta\gamma)$ is specified by Eqs. (1.10). The classical kinetic energy T of the rotator is given by

$$T = \tfrac{1}{2}(I_1\omega_1^2 + I_2\omega_2^2 + I_3\omega_3^2), \tag{3.1}$$

where ω_i is the angular velocity of the object about principal axis i. To relate this to $\dot{\alpha}$, $\dot{\beta}$, and $\dot{\gamma}$, we consider an infinitesimal displacement of the rotator. Since the frame F' is carried with the rotator, we are easily able to find dx, dy, and dz in terms of $d\alpha$, $d\beta$, and $d\gamma$ by taking ξ, η, and ζ to be constant and differentiating the inverses of Eqs. (1.10). If, however, frame F' had not moved with the rotator, the point $(\xi\eta\zeta)$ fixed in the rotator would have been assigned coordinates $(\xi + d\xi,\ \eta + d\eta,\ \zeta + d\zeta)$, and the increments $d\xi$, $d\eta$, and $d\zeta$ can be immediately found by substituting dx for x, etc., in Eqs. (1.10). We get

$$\omega_1 = \left(\frac{\eta\dot{\zeta} - \zeta\dot{\eta}}{\eta^2 + \zeta^2}\right)_{\xi=0} = \dot{\beta}\sin\gamma - \dot{\alpha}\sin\beta\cos\gamma,$$

$$\omega_2 = \left(\frac{\zeta\dot{\xi} - \xi\dot{\zeta}}{\zeta^2 + \xi^2}\right)_{\eta=0} = \dot{\beta}\cos\gamma + \dot{\alpha}\sin\beta\sin\gamma,$$

$$\omega_3 = \left(\frac{\xi\dot{\eta} - \eta\dot{\xi}}{\xi^2 + \eta^2}\right)_{\zeta=0} = \dot{\alpha}\cos\beta + \dot{\gamma}.$$

Readers with a well-developed geometrical perception could no doubt write down these expressions for ω_i without any preliminaries, as do Landau and Lifshitz [32]. The generalized momenta are defined by

$$\begin{aligned} p_\alpha &= \partial T/\partial\dot{\alpha} = -I_1\omega_1\sin\beta\cos\gamma + I_2\omega_2\sin\beta\sin\gamma + I_3\omega_3\cos\beta, \\ p_\beta &= \partial T/\partial\dot{\beta} = I_1\omega_1\sin\gamma + I_2\omega_2\cos\gamma, \\ p_\gamma &= \partial T/\partial\dot{\gamma} = I_3\omega_3. \end{aligned}$$

We now make the replacements $p_\alpha = -i\hbar\partial/\partial\alpha$, etc., and, after having done this, express $I_i\omega_i$ as sums of these derivatives. The result is

$$I_1\omega_1 = -\hbar\lambda_\xi, \qquad I_2\omega_2 = -\hbar\lambda_\eta, \qquad I_3\omega_3 = -\hbar\lambda_\zeta, \tag{3.2}$$

where $\boldsymbol{\lambda}$ is given by Eqs. (1.23). The quantum-mechanical Hamiltonian for the rigid rotator is thus

$$T = \frac{1}{2}\hbar^2\left(\frac{\lambda_\xi^2}{I_1} + \frac{\lambda_\eta^2}{I_2} + \frac{\lambda_\zeta^2}{I_3}\right). \tag{3.3}$$

Of course, we could have appealed to classical mechanics for Eq. (3.3) instead of Eq. (3.1). However, the intervening analysis is important for establishing the negative signs in Eqs. (3.2).

To assign quantum numbers to the eigenfunctions of T, we seek operators that commute with it. Now, we have already determined that $\mathbf{l}$, defined in Eqs. (1.22), commutes with every component of $\boldsymbol{\lambda}$. Hence the eigenvalues $J(J+1)$ and M of the two mutually commuting operators $\mathbf{l}^2$ and l_z can be used to label the eigenfunctions. Of course, $\mathbf{l}^2 = \boldsymbol{\lambda}^2$, and the two quantum numbers J and M correspond to the total angular momentum of the rigid rotator and its z projection in the laboratory frame. The eigenfunction ψ can only depend on the Euler angles ω, since these are the only dynamical variables appearing in the Hamiltonian. As $\mathfrak{D}_{MN}{}^J(\omega)^*$ is an eigenfunction of $\mathbf{l}^2$ and l_z, it is highly convenient to expand $\psi(\omega)$ in terms of the $\mathfrak{D}$ functions. This is possible because the $\mathfrak{D}$ functions (like the spherical harmonics Y_{nlm}) form a complete set. Thus the eigenfunction corresponding to a specified J and M takes the form

$$\psi(\omega) = \sum_N a_N \mathfrak{D}_{MN}{}^J(\omega)^* \tag{3.4}$$

for the rigid rotator.

The coefficients a_N in Eq. (3.4) depend on the moments of inertia, I_i. If two of these (say I_1 and I_2) are equal, the rigid rotator is called a *symmetric top*. For this special case, we can write

$$T = \frac{1}{2}\hbar^2\left[\frac{1}{I_1}\boldsymbol{\lambda}^2 + \left(\frac{1}{I_3} - \frac{1}{I_1}\right)\lambda_\zeta{}^2\right],$$

and this expression for T commutes with λ_ζ as well as with $\mathbf{l}^2$ and l_z. From Section 1.6, we know $\mathfrak{D}_{MN}{}^J(\omega)^*$ is an eigenfunction of λ_ζ with eigenvalue $-N$, so a single $\mathfrak{D}$ function is an eigenfunction of the symmetric top. We impose a normalization condition and write

$$\psi(\omega) = \{(2J+1)/8\pi^2\}^{1/2}\mathfrak{D}_{MN}{}^J(\omega)^* \tag{3.5}$$

Since $I_3\omega_3 = -\hbar\lambda_\zeta$, the angular momentum of the symmetric top about the symmetry axis ζ is $\hbar N$ for the eigenfunction of Eq. (3.5). The eigenvalues of T are

$$\frac{1}{2}\hbar^2\left[\frac{J(J+1)}{I_1} + \left(\frac{1}{I_3} - \frac{1}{I_1}\right)N^2\right], \tag{3.6}$$

and we must necessarily have $J \geq |N|$. Unless $I_1 = I_3$ $(= I_2)$, in which case the symmetric top reduces to a *spherical rotator*, each energy level for which $|N| > 0$ possesses a degeneracy of $2(2J+1)$. The levels of a spherical rotator exhibit a degeneracy of $(2J+1)^2$. Needless to say, any linear combination of the solutions (3.5) that correspond to the same energy is itself a solution.

Although the eigenfunctions $\psi(\omega)$ of Eq. (3.5) form basis functions for an irreducible representation of $R(4)$ if we let M and N run between their limits of $-J$ and J, they correspond to a single energy level only for the spherical rotator. This is because the operators λ_ξ and λ_η do not commute with the Hamiltonian of the symmetric top. The group $R(4)$ is said to be a *noninvariance* group for this system.

3.2 REVERSED ANGULAR MOMENTUM

To convert $\psi(\omega)$ of Eq. (3.5) to the form of a ket, we write

$$\psi(\omega) = \{(2J+1)/8\pi^2\}^{1/2}\mathfrak{D}_{MN}{}^J(\omega)^* = (-1)^{J-N}\,|\,J, M, -N\rangle. \quad (3.7)$$

The reason for the phase factor $(-1)^{J-N}$ and the minus sign preceding N in the ket is precisely the same as the reason for their appearance in Eq. (1.27). Both quantum numbers M and $-N$ are on a similar footing with respect to $\mathbf{l}$ and $\boldsymbol{\lambda}$. Thus M and $-N$ specify the eigenvalues of l_z and λ_ζ, and the properties of the shift operators $l_x \pm il_y$ and $\lambda_\xi \pm i\lambda_\eta$ are also matched.

The presence of $-N$ rather than N in the ket may seem distasteful. As has just been pointed out in the previous section, the angular momentum of the symmetric top about the symmetry axis is $\hbar N$, and not $-\hbar N$. It might therefore seem more natural to write $|\,J, MN\rangle$ for $\psi(\omega)$ and to introduce the vector $\boldsymbol{\lambda}' = -\boldsymbol{\lambda}$, so that N now appears as the eigenvalue of λ_ζ'. Unfortunately, these modest adjustments entail another change. Unlike $\boldsymbol{\lambda}$, the vector $\boldsymbol{\lambda}'$ is not an angular momentum vector, since it satisfies the commutation relations

$$[\lambda_\xi', \lambda_\eta'] = -i\lambda_\zeta', \quad (3.8)$$

etc., in which $-i$ appears on the right-hand side instead of i. However, this is not entirely satisfactory, since $\boldsymbol{\lambda}'$ represents, in one sense, an angular momentum: it is, in fact, the total angular momentum $\mathbf{l}$ of the rigid rotator projected onto axes coinciding instantaneously with those of the frame F' fixed in the rotator. To see this, we write $\mathbf{l} = \mathbf{l}^{(10)}$ in the double-tensor notation of Section 1.7. To project $\mathbf{l}^{(10)}$ onto the axes of F', we have merely to construct $(\mathbf{D}^{(11)}\mathbf{l}^{(10)})^{(01)}$. Using Eqs. (1.22)–(1.23), we find

$$\boldsymbol{\lambda}' = (\mathbf{D}^{(11)}\mathbf{l}^{(10)})^{(01)}. \quad (3.9)$$

(An alternative method to the direct approach for obtaining this result is outlined in Problem 3.2.) Van Vleck [33] realized that matters could be arranged to make it permissible to take Eq. (3.8) as the standard form for the commutation relations for the components of an angular momentum vector. The sign of i can have no significance in quantum mechanics: one merely has to be consistent. However, if we accept Eq. (3.8) and its cyclic permutations as defining an angular momentum vector, a normal angular momentum vector **P** satisfies anomalous commutation relations. To correct this, Van Vleck defined, for such a vector, the reversed angular momentum $\tilde{\mathbf{P}}$ by the equation

$$\tilde{\mathbf{P}} = -\mathbf{P}, \tag{3.10}$$

and now the components of $\tilde{\mathbf{P}}$ satisfy commutation relations of the type (3.8). The kets $| \tilde{P}, \tilde{M}_P \rangle$ are to be used, in which $\tilde{M}_P$ is the eigenvalue of $\tilde{P}_\zeta$.

The change of sign of an angular momentum vector corresponds to time reversal. This can be seen to be associated with the replacement $i \rightarrow -i$ by substituting $-\mathbf{p}$ for $\mathbf{p}$ in the fundamental commutation relations (1.1). Freed [34] has described in detail the procedures to follow when the method of reversed angular momentum is adopted.

It is not necessary to take Van Vleck's extreme position. Carrington *et al.* [35] have pointed out that the commutation relations (3.8) can be corrected by setting

$$\lambda_\xi' = \lambda_\xi'', \qquad \lambda_\eta' = -\lambda_\eta'', \qquad \lambda_\zeta' = \lambda_\zeta'',$$

for then $\boldsymbol{\lambda}''$ satisfies the commutation relations of an ordinary angular momentum vector.

Alternatively, all these adjustments can be avoided by accepting Eq. (3.7), and this is the course we shall follow here. The price we pay is a certain asymmetry between the frames F and F' to which our double tensors are referred (see Problem 3.2).

3.3 REDUCED MATRIX ELEMENTS

An important matrix element for eigenstates of the rigid rotator is

$$\langle JMN \mid D_{pq}^{(kk)} \mid J'M'N' \rangle. \tag{3.11}$$

The notation of Eq. (3.7) is followed in bra and ket. (Thus N is the eigenvalue of λ_ζ.) From this equation and from Eq. (1.27), the matrix element (3.11) is equal to

$$(-1)^{J+N+J'+N'+k+q}[J, k, J']^{1/2} \frac{1}{8\pi^2} \int \mathfrak{D}_{M-N}{}^{J}(\omega)\mathfrak{D}_{p-q}{}^{k}(\omega)^* \mathfrak{D}_{M'-N'}{}^{J'}(\omega)^* \, d\omega.$$

The integral can be readily evaluated from Eqs. (1.14) and (1.45).

On the other hand, the WE theorem can be applied to the matrix element (3.11). Since we are dealing with a double tensor, two 3-j symbols appear, as in Eq. (1.58). They exactly match those coming from the evaluation of the integral, and we get

$$(J \,||\, D^{(kk)} \,||\, J') = [J, k, J']^{1/2}. \tag{3.12}$$

Strictly speaking, the symbols J and J' in the bra and ket should each be repeated, to indicate that the matrix element is reduced in two spaces. The abbreviation should not lead to confusion, however.

As will be seen later, we often need matrix elements that are reduced in one of the two spaces, but not the other. For a matrix element reduced in the M space only (i.e., the space of the laboratory frame F), we have

$$(JN \,|\!:\! D_{\cdot q}^{(kk)} :\!|\, J'N') = (-1)^{J-N} \begin{pmatrix} J & k & J' \\ -N & q & N' \end{pmatrix} [J, k, J']^{1/2}. \tag{3.13}$$

For a reduced matrix element in the N space (i.e., the space of the frame F' fixed in the rotator), we have, in an analogous way,

$$(JM :\!|\, D_{p\cdot}^{(kk)} \,|\!:\! J'M') = (-1)^{J-M} \begin{pmatrix} J & k & J' \\ -M & p & M' \end{pmatrix} [J, k, J']^{1/2}. \tag{3.14}$$

The dots subscripted to D merely indicate the blank spaces produced by the reduction process.

3.4 MOMENTUM SPACE

It has been seen that the rotation matrices $\mathfrak{D}$ serve as eigenfunctions for the symmetric top. The four-dimensional spherical harmonics Y_{nlm} also occur in the expression for the eigenfunctions of a simple physical system, namely, the hydrogen atom. For the Y functions to play this role, the analysis must be carried out in momentum space. Discussions of momentum space are given in many texts on quantum mechanics, and it seems unnecessary to do more than sketch a brief introduction here.

To begin with, we notice that the fundamental commutation relations between $\mathbf{p}$ and $\mathbf{r}$ are as well preserved by the substitutions

$$x = i\frac{\partial}{\partial p_x},$$

etc., as by the more familiar replacements

$$p_x = -i\frac{\partial}{\partial x},$$

etc. In fact, there is a complete reciprocity between these two points of view. The expansion of an eigenfunction $\psi(\mathbf{p})$ of momentum space in terms of the eigenfunction $u(\mathbf{r})$ of position space is given by

$$\psi(\mathbf{p}) = (2\pi)^{-3/2} \int \exp(-i\mathbf{r}\cdot\mathbf{p})\, u(\mathbf{r})\, d\mathbf{r}, \tag{3.15}$$

where the integration is carried out over all space. (This is always implied in equations of this type when no limits are given.) Thus $\psi(\mathbf{p})$ is the Fourier transform of $u(\mathbf{r})$. It is supposed that both ψ and u are normalized. (Perhaps it is also worth recalling from Section 1.1 that $\mathbf{p}$ is the linear momentum divided by $\hbar$.) A central formula [6] for carrying out manipulations involving Fourier transforms is

$$\int \exp(i\mathbf{r}\cdot\mathbf{p})\, d\mathbf{p} = (2\pi)^3\delta(\mathbf{r}), \tag{3.16}$$

where the delta function possesses the property, for any continuous function $f(\mathbf{r})$,

$$\int f(\mathbf{r})\delta(\mathbf{r} - \mathbf{r}')\, d\mathbf{r} = f(\mathbf{r}'). \tag{3.17}$$

With the aid of Eqs. (3.16)–(3.17), we may readily obtain the reciprocal relation to Eq. (3.15):

$$u(\mathbf{r}) = (2\pi)^{-3/2} \int \exp(i\mathbf{r}\cdot\mathbf{p})\, \psi(\mathbf{p})\, d\mathbf{p}. \tag{3.18}$$

Suppose $u(\mathbf{r})$ is the eigenfunction for an electron (of mass m) moving in a potential $+V(r)/e$, where $-e$ is the charge of the electron. The Schrödinger equation for this system is

$$[\hbar^2\mathbf{p}^2/2m + V(\mathbf{r})]u(\mathbf{r}) = Eu(\mathbf{r}).$$

To obtain an equation involving $\psi(\mathbf{p})$, we multiply by $\exp(-i\mathbf{r}\cdot\mathbf{p})$ and integrate over $\mathbf{r}$:

$$\begin{aligned}(2\pi)^{3/2}[\hbar^2\mathbf{p}^2/2m - E]\psi(\mathbf{p}) &= -\int V(\mathbf{r})u(\mathbf{r}) \exp(-i\mathbf{r}\cdot\mathbf{p})\, d\mathbf{r} \\ &= -\int V(\mathbf{r}) \exp(-i\mathbf{r}\cdot\mathbf{p})\, (2\pi)^{-3/2} \int \psi(\mathbf{p}') \exp(i\mathbf{r}\cdot\mathbf{p}')\, d\mathbf{p}'\, d\mathbf{r} \\ &= -\int \psi(\mathbf{p}') V'(\mathbf{p} - \mathbf{p}')\, d\mathbf{p}',\end{aligned}$$

where

$$V'(\mathbf{p}'') = (2\pi)^{-3/2} \int d\mathbf{r} V(\mathbf{r}) \exp(-i\mathbf{r}\cdot\mathbf{p}''). \tag{3.19}$$

So the Schrödinger equation becomes an integral equation in $\psi(\mathbf{p})$:

$$(\hbar^2\mathbf{p}^2/2m - E)\psi(\mathbf{p}) = -(2\pi)^{-3/2} \int \psi(\mathbf{p}') V'(\mathbf{p} - \mathbf{p}')\, d\mathbf{p}'. \tag{3.20}$$

3.5 THE COULOMB POTENTIAL

A special case of $V(\mathbf{r})$ which is of great importance to us is

$$V(\mathbf{r}) = -\frac{e^2Z}{|\mathbf{r} - \mathbf{R}|},$$

corresponding to the electron moving in the presence of an electric charge $+eZ$ at the point $\mathbf{R}$. We set $\mathbf{r} - \mathbf{R} = \mathbf{r}'$ and obtain, from Eq. (3.19),

$$V'(\mathbf{p}) = -(2\pi)^{-3/2} e^2 Z \exp(-i\mathbf{R}\cdot\mathbf{p}) \int r'^{-1} \exp(-i\mathbf{r}'\cdot\mathbf{p})\, d\mathbf{r}'.$$

To evaluate the integral, we regard $\mathbf{p}$ (which is constant during the integration) as defining a new z axis (say z') with respect to which the spherical polar coordinate θ' is defined. Then

$$\int r'^{-1} \exp(-i\mathbf{r}'\cdot\mathbf{p})\, d\mathbf{r}' = \int_0^\infty \int_{-1}^1 \int_0^{2\pi} r'\, dr'\, d(\cos\theta')\, d\phi' \exp(-ir'p\cos\theta')$$
$$= 2\pi p^{-2}[-2\cos(r'p)]_0^\infty.$$

The value of $\cos(r'p)$ for $r' \to \infty$ oscillates between $+1$ and -1, and so the integral is indeterminate.

To avoid this difficulty, the Coulomb potential r^{-1} is replaced by $e^{-\alpha r}/r$, where α is small but finite and positive. We find

$$\int r'^{-1} \exp(-i\mathbf{r}'\cdot\mathbf{p} - \alpha r')\, d\mathbf{r}' = 4\pi/(p^2 + \alpha^2),$$

which tends to the determinate value $4\pi/p^2$ as $\alpha \to 0$. If this device seems arbitrary, one can always reflect that α could be chosen sufficiently small to elude detection in any experimental test of Coulomb's law. Thus we get

$$V'(\mathbf{p}) = -4\pi e^2 Z p^{-2} (2\pi)^{3/2} \exp(-i\mathbf{R}\cdot\mathbf{p}). \tag{3.21}$$

Before writing out again the integral form of Schrödinger's equation, we make the substitution

$$E = -\hbar^2 p_0^2/2m, \tag{3.22}$$

getting

$$(\mathbf{p}^2 + p_0^2)\psi(\mathbf{p}) = \frac{Z}{\pi^2 a_0} \int \exp(-i\mathbf{R}\cdot(\mathbf{p} - \mathbf{p}'))\, \psi(\mathbf{p}') \frac{d\mathbf{p}'}{|\mathbf{p} - \mathbf{p}'|^2}, \tag{3.23}$$

where $a_0 = \hbar^2/e^2 m$, the Bohr radius.

3.6 FOUR-DIMENSIONAL MOMENTUM SPACE

To project $\mathbf{p}$ onto the surface of a four-dimensional unit sphere, we write

$$\begin{aligned} x_1 &= 2p_0 p_x (p^2 + p_0^2)^{-1} &&= \sin\chi \sin\theta \cos\phi, \\ x_2 &= 2p_0 p_y (p^2 + p_0^2)^{-1} &&= \sin\chi \sin\theta \sin\phi, \\ x_3 &= 2p_0 p_z (p^2 + p_0^2)^{-1} &&= \sin\chi \cos\theta, \\ x_4 &= (p_0^2 - p^2)(p^2 + p_0^2)^{-1} &&= \cos\chi. \end{aligned} \tag{3.24}$$

Apart from the interchange $p_0^2 \leftrightarrow p^2$ in the definition of x_4, this transforma-

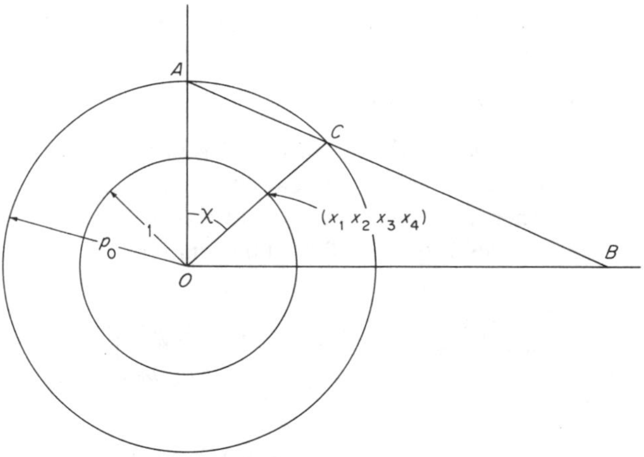

Fig. 3.1. Projection of **p** onto the surface of the unit hypersphere. For illustrative purposes, the vector **p** (represented by OB) is confined to a two-dimensional space, namely a plane perpendicular to the plane of the diagram. The line OA is of length p_0 and is perpendicular to OB. The intersection at C of AB with the sphere centered at O of radius p_0 defines a vector OC which intersects the unit sphere at $(x_1x_2x_3x_4)$. Fock's definition

$$x_4 = (p^2 - p_0^2)/(p^2 + p_0^2) = \cos\chi$$

has been used in constructing this diagram.

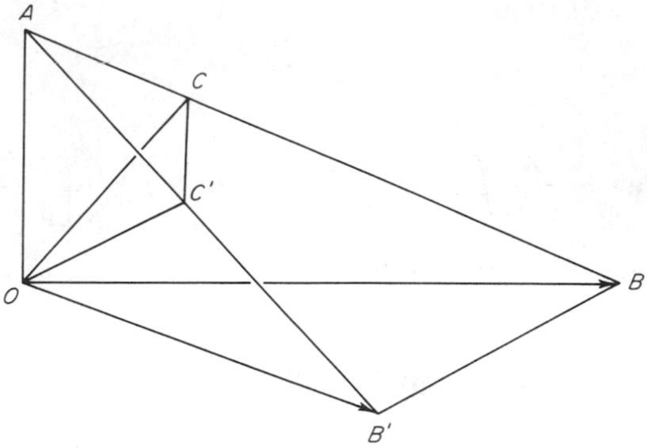

Fig. 3.2. Two momentum vectors, OB and OB', are projected onto the sphere of radius p_0 and center at O by finding the intersections C and C' with the sphere of the lines AB and AB'. The intersections of OC and OC' with the unit sphere centered at O define the points $(x_1x_2x_3x_4)$ and $(x_1'x_2'x_3'x_4')$.

tion is the same as that used by Fock [36] in his celebrated treatment of the hydrogen atom. The altered sign of x_4, which ensures that the limits π and 0 for χ correspond to the respective limits ∞ and 0 for p, has been used by Shibuya and Wulfman [37]. Fock's scheme is particularly suitable for a geometrical interpretation, as can be seen in Fig. 3.1. In either case, we notice that

$$p_x = p \sin\theta \cos\phi, \qquad p_y = p \sin\theta \sin\phi, \qquad p_z = p \cos\theta, \qquad (3.25)$$

so the angles θ and ϕ specify the direction of $\mathbf{p}$ in the usual way.

The quadratic form $|\mathbf{p} - \mathbf{p}'|^2$ appears in Eq. (3.23). To express it in terms of the distance between two points $(x_1 \cdots x_4)$ and $(x_1' \cdots x_4')$ on the hypersphere, we refer to Fig. 3.2. The triangles ACC' and $AB'B$ are similar, since

$$\frac{AC'/2}{AC/2} = \frac{p_0 \sin \chi'/2}{p_0 \sin \chi/2} = \left(\frac{p^2 + p_0^2}{p'^2 + p_0^2}\right)^{1/2},$$

whereas

$$\frac{AB}{AB'} = \frac{(p^2 + p_0^2)^{1/2}}{(p'^2 + p_0^2)^{1/2}}.$$

Hence

$$CC'/BB' = AC/AB',$$

which yields

$$\frac{CC'}{|\mathbf{p} - \mathbf{p}'|} = \frac{2p_0^2}{(p^2 + p_0^2)^{1/2}(p'^2 + p_0^2)^{1/2}}.$$

Now

$$CC' = p_0\{\sum (x_a - x_a')^2\}^{1/2},$$

so we obtain the required result,

$$\sum_{a=1}^{4} (x_a - x_a')^2 = \frac{4p_0^2 \, | \mathbf{p} - \mathbf{p}' |^2}{(p^2 + p_0^2)(p'^2 + p_0^2)} \,. \tag{3.26}$$

If κ is the angle between OC and OC', we have

$$\cos \kappa = \sum_{a=1}^{4} x_a x_a'$$

from Eq. (2.41), so that

$$\sum_{a=1}^{4} (x_a - x_a')^2 = 2 - 2\cos\kappa = 4 \sin^2 \kappa/2.$$

Thus we can write

$$\frac{1}{| \mathbf{p} - \mathbf{p}' |^2} = \frac{p_0^2}{(p^2 + p_0^2)(p'^2 + p_0^2) \sin^2 \kappa/2} \,. \tag{3.27}$$

Before the integral form of Schrödinger's equation can be used, we need to relate the infinitesimal volume $d\mathbf{p}$ to the surface element $d\Omega$ of the hypersphere. Comparison of

$$d\mathbf{p} = -p^2 \, dp \, d(\cos\theta) \, d\phi$$

with

$$d\Omega = -\sin^2 \chi \, d\chi \, d(\cos\theta) \, d\phi,$$

shows that

$$\frac{d\mathbf{p}}{d\Omega} = \frac{p^2}{\sin^2 \chi} \frac{dp}{d\chi} \,,$$

and we at once obtain

$$\frac{dp}{d\chi} = \frac{p^2 + p_0^2}{2p_0}$$

from the last equation of the set (3.24). Thus

$$d\mathbf{p} = \left(\frac{p^2 + p_0^2}{2p_0}\right)^3 d\Omega. \tag{3.28}$$

Equations (3.27)–(3.28) can now be used to replace $| \mathbf{p} - \mathbf{p}' |^{-2}$ and $d\mathbf{p}'$ in Eq. (3.23). The resulting equation is somewhat encumbered by terms in $(p^2 + p_0^2)$ and $(p'^2 + p_0^2)$. These are removed by taking

$$\psi(\mathbf{p}) \equiv \frac{4p_0^{5/2}}{(p^2 + p_0^2)^2}\varphi(\Omega), \tag{3.29}$$

the term $4p_0^{5/2}$ being included in anticipation of normalization requirements for the hydrogen atom. Eq. (3.23) now takes the remarkably simple form,

$$\varphi(\Omega) = \frac{Z}{2\pi^2 a_0 p_0}\int \frac{\exp(-i\mathbf{R}\cdot(\mathbf{p}-\mathbf{p}'))}{4\sin^2\kappa/2}\varphi(\Omega')\,d\Omega'. \tag{3.30}$$

3.7 THE HYDROGEN ATOM

Equation (3.30) is an example of a Fredholm equation of the second kind [38]. The general method for solving such an equation is to expand both kernel and function in some series of orthonormal polynomials, thereby obtaining a series of equations for the coefficients in the expansion of the function. The natural choice for our polynomials is the four-dimensional spherical harmonics. If we specialize to the hydrogen atom, we can set $\mathbf{R} = 0$, corresponding to a charge Ze at the origin. To expand the kernel $\csc^2\kappa/2$, we can use Eq. (2.44), getting

$$\varphi(\Omega) = \frac{Z}{a_0 p_0}\sum_{nlm}\frac{1}{n}\int Y_{nlm}^*(\Omega')Y_{nlm}(\Omega)\varphi(\Omega')\,d\Omega'.$$

We now develop $\varphi(\Omega)$ as a series of spherical harmonics:

$$\varphi(\Omega) = \sum_{\nu\lambda\mu} a_{\nu\lambda\mu}Y_{\nu\lambda\mu}(\Omega).$$

The integration over Ω' can be performed at once, and we get

$$\sum_{\nu\lambda\mu} a_{\nu\lambda\mu}Y_{\nu\lambda\mu}(\Omega) = \frac{Z}{a_0 p_0}\sum_{nlm}\frac{1}{n}a_{nlm}Y_{nlm}(\Omega).$$

To obtain equations for the coefficients $a_{\nu\lambda\mu}$, both sides of this equation are multiplied by $Y_{n'l'm'}^*(\Omega)$ and the integration over Ω is carried out. The resulting secular matrix is diagonal, and for its determinant to vanish we must have

$$1 = Z/a_0 p_0 n$$

for some positive integer n. Thus an acceptable energy is

$$E = -\frac{\hbar^2}{2m}p_0^2 = -\frac{\hbar^2 Z^2}{2ma_0^2 n^2} = -\frac{e^4 m Z^2}{2\hbar^2 n^2},$$

and we recover the familiar expression for the energies of the levels of the nonrelativistic hydrogen atom. Since the secular matrix is diagonal, we have

$$\varphi(\Omega) = Y_{nlm}(\Omega).$$

This means that a solution for the momentum-space eigenfunction $\psi(\mathbf{p})$ is given by

$$\psi(\mathbf{p}) = 4p_0^{5/2}(p^2 + p_0^2)^{-2}Y_{nlm}(\Omega). \tag{3.31}$$

The subscripts n, l, and m now play their traditional spectroscopic roles as the principal, azimuthal, and magnetic quantum numbers respectively.

Explicit expressions for $\psi(\mathbf{p})$ were first obtained for the hydrogen atom by Podolsky and Pauling [39], who directly calculated the Fourier transforms of the standard position-space eigenfunctions $u(\mathbf{r})$. The latter are given by

$$u(\mathbf{r}) = 2\left(\frac{Z}{na_0}\right)^{3/2}\left[\frac{(n-l-1)!}{n(n+l)!}\right]^{1/2} e^{-w/2}w^l L_{n-l-1}{}^{2l+1}(w)\, Y_{lm}(\theta, \phi), \tag{3.32}$$

where $w = 2Zr/na_0$. The Laguerre polynomials L are defined by

$$L_\alpha{}^\beta(x) = \frac{e^x}{\alpha!x^\beta}\frac{d^\alpha}{dx^\alpha}(x^{\alpha+\beta}e^{-x}). \tag{3.33}$$

The phase in Eq. (3.31) has been chosen to ensure that Eqs. (3.15), (3.31) and (3.32) are consistent.

3.8 THE LENZ VECTOR

In the previous chapter, we examined in some detail the properties of the four-dimensional spherical harmonics $Y_{nlm}(\Omega)$. Now that Eq. (3.31) has been established, it becomes interesting to see what physical significance can be attached to generators of the group $R(4)$ associated with these harmonics. The connection between mathematics and physics is made in Eqs. (3.24). The generators $\mathbf{L}$ and $\mathbf{A}$ have already been converted to functions of χ, θ, and ϕ in Section 2.4, and we can use Eqs. (3.25), augmented by

$$p = p_0 \tan \chi/2 \tag{3.34}$$

to express these functions in terms of $\mathbf{p}$ and $\partial/\partial\mathbf{p}$. The latter can be removed by making the replacement $\partial/\partial\mathbf{p} \to -i\mathbf{r}$. Thus, in the notation of Section 2.4,

$$L_z = l_z = \frac{1}{i}\frac{\partial}{\partial\phi} = -\left(\frac{\partial p_x}{\partial\phi}x + \frac{\partial p_y}{\partial\phi}y + \frac{\partial p_z}{\partial\phi}z\right)$$

$$= xp_y - yp_x.$$

Other components of $\boldsymbol{l}$ can be handled in a similar way, and we at once see that $\boldsymbol{l}$ is the angular momentum of the electron—a result which the notation has prepared us for.

The other generator, $\mathbf{A}$, is rather more interesting. A similar analysis to that just performed for $\boldsymbol{l}$ yields

$$\mathbf{A} = [-(\mathbf{p} \times \boldsymbol{l}) + \tfrac{1}{2}(p^2 + p_0^2)\mathbf{r}]/p_0. \tag{3.35}$$

As in the case of $\boldsymbol{l}$, replacements of the type $\partial/\partial p_x \to -ix$ have been made to dispose of derivatives with respect to $\mathbf{p}$. The components of $\mathbf{A}$, when acting on the harmonics $Y_{nlm}(\Omega)$, cannot change n, since the collection of harmonics of a given n form bases for an irreducible representation of $R(4)$. (See, for example, Eq. (2.25).) However, it does not follow that this is true when $\mathbf{A}$ acts on the complete eigenfunction $\psi(\mathbf{p})$, since $\mathbf{A}$, unlike $\boldsymbol{l}$, does not commute with the prefacing function $4p_0^{5/2}(p^2 + p_0^2)^{-2}$. To correct this, we define

$$\mathbf{a} = \mathbf{A} + 2i\mathbf{p}/p_0. \tag{3.36}$$

The new vector $\mathbf{a}$ satisfies

$$\mathbf{a}\psi(\mathbf{p}) = \frac{4p_0^{5/2}}{(p^2 + p_0^2)^2}\mathbf{A}Y_{nlm}(\Omega),$$

and hence can only produce new eigenfunctions $\psi(\mathbf{p})$ with the same n. Using Eqs. (3.35)–(3.36), we get

$$\mathbf{a} = [(\boldsymbol{l} \times \mathbf{p}) + \tfrac{1}{2}(p^2 + p_0^2)\mathbf{r}]/p_0$$
$$= [(\boldsymbol{l} \times \mathbf{p}) - (\mathbf{p} \times \boldsymbol{l}) + \mathbf{r}(p^2 + p_0^2)]/2p_0.$$

Now $(p^2 + p_0^2)$, when acting on an eigenfunction $\psi(\mathbf{p})$, is equivalent to $2Z/ra_0$. This follows because the Hamiltonian H for the hydrogen atom is given by

$$H = \hbar^2p^2/2m - Ze^2/r,$$

and the eigenvalue E is $-\hbar^2p_0^2/2m$. Thus we have

$$\mathbf{a} = [(\boldsymbol{l} \times \mathbf{p}) - (\mathbf{p} \times \boldsymbol{l}) + 2Z\mathbf{r}/ra_0]/2p_0. \tag{3.37}$$

This is the quantum-mechanical form of the so-called *Lenz* (or *Runge–*

Lenz) vector. Classically it defines the direction of the major axis of an elliptical orbit. Like the angular momentum vector, it is of constant magnitude (see Problem 3.5). In classical mechanics, as well as in quantum mechanics, $\boldsymbol{l}\cdot\mathbf{a} = 0$. Both $\boldsymbol{l}$ and $\mathbf{a}$ commute with the Hamiltonian and are thus constants of the motion. The quantum-mechanical form of $\mathbf{a}$ was first introduced by Pauli [40].

The Lenz vector shares many properties with the generator $\mathbf{A}$. In analogy with Eqs. (2.3), we can define two vectors $\mathbf{k}_1$ and $\mathbf{k}_2$ by writing

$$\mathbf{k}_1 = \tfrac{1}{2}(\boldsymbol{l} + \mathbf{a}), \qquad \mathbf{k}_2 = \tfrac{1}{2}(\boldsymbol{l} - \mathbf{a}). \tag{3.38}$$

It is at once found that $\mathbf{k}_1$ and $\mathbf{k}_2$ are angular momentum vectors, and also that

$$[\mathbf{k}_1, \mathbf{k}_2] = 0. \tag{3.39}$$

This means that $\mathbf{k}_1$ and $\mathbf{k}_2$, like $\mathbf{K}_1$ and $\mathbf{K}_2$, can be taken as the generators of an $R(4)$ group. Both $\boldsymbol{l}$ and $\mathbf{a}$ commute with the Hamiltonian, as we have just pointed out; and hence $\mathbf{k}_1$ and $\mathbf{k}_2$ do too (see Problem 3.6). The eigenfunctions corresponding to a given energy level form basis functions for an irreducible representation of the new $R(4)$ group. It was shown in Section 2.2 that such representations are the product of two identical irreducible representations of $R(3)$: their dimensions must therefore be of the form $(2k + 1)^2$, where $k = 0, \frac{1}{2}, 1, \ldots$. This sequence of numbers precisely corresponds to the orbital degeneracies of the levels $n = 1, 2, 3, \ldots$ of the nonrelativistic hydrogen atom.

PROBLEMS

3.1 A series of equal point masses are located on the vertices of a regular n-sided polygon. Prove that, for $n > 2$, the moment of inertia of this classical system about any axis lying in the plane of the polygon and passing through the center of the polygon is the same. Deduce that systems of masses exhibiting at least a threefold axial symmetry are symmetric tops.

3.2 In the notation of Section 1.7, we write

$$\boldsymbol{l}^{(01)} = (\mathbf{D}^{(11)}\boldsymbol{l}^{(10)})^{(01)},$$

where $\boldsymbol{l}^{(10)}$ is defined as in Eqs. (1.22). Use the result of Problem 1.6, extended to double tensors, to express $(\mathbf{D}^{(11)}\boldsymbol{l}^{(10)})^{(01)}$ as a product of reduced matrix elements of $\mathbf{D}^{(11)}$ and $\boldsymbol{l}^{(10)}$. Prove

$$(J \,\|\, l^{(01)} \,\|\, J) = -(J \,\|\, l^{(10)} \,\|\, J) = -(2J + 1)\{J(J + 1)\}^{1/2}.$$

Hence obtain the result

$$\boldsymbol{\lambda}' = -\boldsymbol{\lambda} = \boldsymbol{l}^{(01)}$$

in the notation of Section 3.2. Check that $\boldsymbol{l}^{(10)}$ is a vector with respect to $\boldsymbol{l}$ and that $\boldsymbol{l}^{(01)}$ is a vector with respect to $\boldsymbol{\lambda}$. Verify that the components of $\boldsymbol{l}^{(10)}$ satisfy the same commutation relations with respect to one another as those satisfied by an angular momentum vector, but that an identical statement cannot be made for $\boldsymbol{l}^{(01)}$.

3.3 Derive Eq. (3.26) algebraically.

3.4 By taking the Fourier transform of

$$u_{1s}(\mathbf{r}) = \pi^{-1/2}(Z/a_0)^{3/2}\exp(-Zr/a_0),$$

prove that the momentum-space eigenfunction for the ground state of the hydrogen atom is

$$\psi_{1s}(\mathbf{p}) = 2(2)^{1/2}\pi^{-1}(a_0/Z)^{3/2}\{1 + (a_0 p/Z)^2\}^{-2}.$$

3.5 A particle of mass m and charge $-e$ describes a classical elliptic orbit of the type

$$r = (A + B\cos\theta)^{-1}$$

under the influence of a charge $+e$ at the origin. Prove that the classical angular momentum l is given by

$$Al^2 = e^2 m,$$

and that the Lenz vector

$$\mathbf{p} \times \boldsymbol{l} - me^2\mathbf{r}/r,$$

where $\mathbf{p}$ is the momentum of the particle, is of constant magnitude Bl^2 and is directed along the major axis of the ellipse.

3.6 Prove that the angular momentum $\boldsymbol{l}$ commutes with the Hamiltonian H of the nonrelativistic hydrogen atom, and obtain the commutation relations

$$\left[\frac{2Z\mathbf{r}}{ra_0}, \frac{\hbar^2 p^2}{2m}\right] = 2Ze^2\left\{-\frac{\mathbf{r}}{r} + \frac{i}{r}\mathbf{p} - \frac{i}{r^3}\mathbf{r}(\mathbf{r}\cdot\mathbf{p})\right\},$$

$$\left[\boldsymbol{l} \times \mathbf{p} - \mathbf{p} \times \boldsymbol{l}, -\frac{Ze^2}{r}\right] = 2Ze^2\left\{\frac{\mathbf{r}}{r} - \frac{i}{r}\mathbf{p} + \frac{i}{r^3}\mathbf{r}(\mathbf{r}\cdot\mathbf{p})\right\}.$$

Deduce that the Lenz vector **a** commutes with H. Derive the equations

$$(\boldsymbol{l} \times \mathbf{p})^2 = \boldsymbol{l}^2\mathbf{p}^2,$$

$$(\boldsymbol{l} \times \mathbf{p} - \mathbf{p} \times \boldsymbol{l})^2 = 4(\boldsymbol{l}^2 + 1)\mathbf{p}^2,$$

$$(\boldsymbol{l} \times \mathbf{p} - \mathbf{p} \times \boldsymbol{l})\cdot \mathbf{r}r^{-1} + r^{-1}\mathbf{r}\cdot(\boldsymbol{l} \times \mathbf{p} - \mathbf{p} \times \boldsymbol{l}) = -4(\boldsymbol{l}^2 + 1)r^{-1},$$

and hence show that we may write

$$H = -\left(\frac{Z^2e^2}{2a_0}\right)\frac{1}{\mathbf{a}^2 + \boldsymbol{l}^2 + 1}.$$

Deduce that the eigenvalues E of H are given by

$$E = -\left(\frac{Z^2e^2}{2a_0}\right)\frac{1}{(2k + 1)^2},$$

where $k = 0, \frac{1}{2}, 1, \ldots.$

4

The Hydrogen Molecular Ion

4.1 INTRODUCTION

The hydrogen molecular ion H_2^+, comprising two protons and one electron, is the simplest molecule of all. If we neglect all effects coming from the spins of the particles, and if, in addition, the motion of the protons is disregarded, we are faced with the problem of an electron moving in the presence of two fixed Coulomb centers of attraction. In this form, the analysis of H_2^+ has attracted considerable attention over the years. One of the first treatments was given in terms of the old quantum theory by Pauli in his Munich thesis [41]: reviews of subsequent developments as well as general accounts of modern procedures for getting accurate eigenfunctions have been given by Buckingham [42] and by Teller and Sahlin [43]. The separability of Schrödinger's equation in spheroidal coordinates enables solutions of almost unlimited accuracy to be computed. The data of Teller and Sahlin augment the earlier work of Bates and his collaborators [44, 45]. Additional numerical results have been provided by Sharp [46] and by Madsen and Peek [47]. Extensions to the heteronuclear case have been studied by Wilson and Gallup [48], and by Hunter and Pritchard [49], among others. The relativistic analysis of H_2^+ by Pavlik and Blinder [50] should also be mentioned.

The methods for finding accurate numerical eigenfunctions will not be described here. Rather, we shall emphasize the structural aspects of the analysis. In particular, we shall examine possible applications of angular momentum theory, as well as the simplifications that occur for special ratios of the charges on the two centers.

4.2 SPHEROIDAL COORDINATES

The spheroidal (i.e., confocal ellipsoidal) coordinate scheme is the natural one to use for a two-center system. Although these coordinates are treated in many texts, it seems appropriate to at least summarize their properties here. To begin with, the symbols ξ and η are released from their previous roles in readiness for their use as two of the three spheroidal coordinates $(\xi\eta\phi)$ of a point X. (This notational change will be maintained until the beginning of Chapter 6.) If the point X is distant r_A and r_B from the two centers A and B respectively, we write

$$\xi = (r_A + r_B)/2b, \qquad \eta = (r_A - r_B)/2b, \tag{4.1}$$

where $2b$ is the separation AB (see Fig. 4.1). The symbol ϕ defines the angle between the plane ABX and a fixed plane through A and B. Lines of constant $(\xi\eta)$, $(\xi\phi)$, and $(\eta\phi)$ are circles, ellipses, and hyperbolas respectively. They intersect at right angles, and all ellipses and hyperbolas possess A and B as foci. The connection between $(\xi\eta\phi)$ and a Cartesian

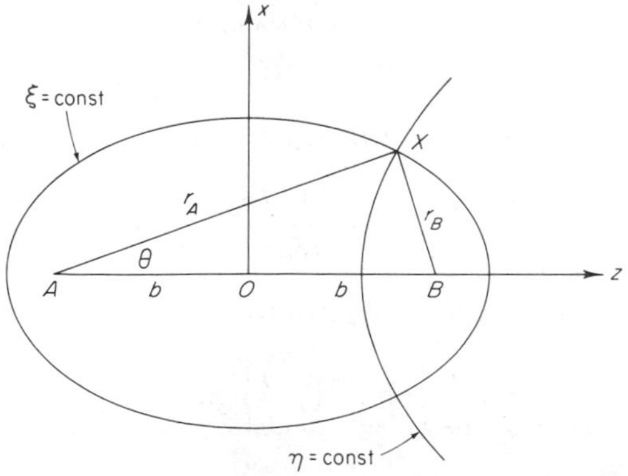

Fig. 4.1. Diagram for spheroidal coordinates. To simplify the figure, the point X has been drawn for the special case for which $\phi = 0$, corresponding to the coincidence of the xz plane and the plane ABX.

coordinate scheme, in which the origin O is taken as the midpoint of AB, with OB being the z axis, is

$$x = b\{(\xi^2 - 1)(1 - \eta^2)\}^{1/2} \cos\phi,$$
$$y = b\{(\xi^2 - 1)(1 - \eta^2)\}^{1/2} \sin\phi,$$
$$z = b\xi\eta. \tag{4.2}$$

It is sometimes convenient to take A as the origin; the new Cartesian coordinates $(x'y'z')$ satisfy

$$x' = x, \qquad y' = y, \qquad z' = z + b. \tag{4.3}$$

The angle θ between AB and AX is given by

$$\cos\theta = \frac{\xi\eta + 1}{\xi + \eta}, \qquad \sin\theta = \frac{\{(\xi^2 - 1)(1 - \eta^2)\}^{1/2}}{\xi + \eta}. \tag{4.4}$$

The entire space is covered by allowing ξ, η, and ϕ to run over the regions defined by

$$1 \le \xi \le \infty, \qquad -1 \le \eta \le 1, \qquad 0 \le \phi \le 2\pi. \tag{4.5}$$

The interval ds is given by

$$ds^2 = h_1^2\, d\xi^2 + h_2^2\, d\eta^2 + h_3^2\, d\phi^2,$$

in which

$$h_1 = b\sqrt{\frac{\xi^2 - \eta^2}{\xi^2 - 1}}, \qquad h_2 = b\sqrt{\frac{\xi^2 - \eta^2}{1 - \eta^2}},$$
$$h_3 = b\{(\xi^2 - 1)(1 - \eta^2)\}^{1/2}, \tag{4.6}$$

and so the infinitesimal volume element $d\tau$ is given by

$$d\tau = h_1\, d\xi\, h_2\, d\eta\, h_3\, d\phi$$
$$= b^3(\xi^2 - \eta^2)\, d\xi\, d\eta\, d\phi. \tag{4.7}$$

For the general orthogonal coordinate system $(x_1x_2x_3)$, the Laplacian is

$$\nabla^2 = \frac{1}{h_1h_2h_3}\left[\frac{\partial}{\partial x_1}\left(\frac{h_2h_3}{h_1}\frac{\partial}{\partial x_1}\right) + \frac{\partial}{\partial x_2}\left(\frac{h_3h_1}{h_2}\frac{\partial}{\partial x_2}\right) + \frac{\partial}{\partial x_3}\left(\frac{h_1h_2}{h_3}\frac{\partial}{\partial x_3}\right)\right].$$

When $(x_1x_2x_3) \equiv (\xi\eta\phi)$, this evaluates to

$$\nabla^2 = \frac{b^{-2}}{\xi^2 - \eta^2}\left[\frac{\partial}{\partial\xi}\left\{(\xi^2 - 1)\frac{\partial}{\partial\xi}\right\} + \frac{\partial}{\partial\eta}\left\{(1 - \eta^2)\frac{\partial}{\partial\eta}\right\}\right.$$
$$\left. + \frac{\xi^2 - \eta^2}{(\xi^2 - 1)(1 - \eta^2)}\frac{\partial^2}{\partial\phi^2}\right]. \tag{4.8}$$

The transformation (4.2) allows us to express the angular momentum vector $\boldsymbol{l}$ (defined in Eqs. (1.6)) in terms of ξ, η, and ϕ. The result is

$$l_\pm = \pm e^{\pm i\phi} \frac{\{(\xi^2 - 1)(1 - \eta^2)\}^{1/2}}{\xi^2 - \eta^2} \left(\eta \frac{\partial}{\partial \xi} - \xi \frac{\partial}{\partial \eta}\right)$$

$$+ e^{\pm i\phi} \frac{i\xi\eta}{\{(\xi^2 - 1)(1 - \eta^2)\}^{1/2}} \frac{\partial}{\partial \phi},$$

$$l_z = \frac{1}{i} \frac{\partial}{\partial \phi}. \tag{4.9}$$

Spheroidal coordinates are the appropriate ones to use when analyzing the classical motion of a particle under the gravitational attraction of two fixed centers. Unlike the motion around a single center, the general orbit is not a closed curve, but rather a complicated trajectory that can only be usefully described in general terms. However, the limits of the region swept out by a particular trajectory are well defined. If ϕ is constant, corresponding to motion in a plane, the boundaries are always some combination of the ellipses and hyperbolas for which ξ and η are constant respectively. These boundaries go over into ellipses and hyperbolas of revolution when the particle possesses nonzero angular momentum about the axis AB. The particular combination of ellipses and hyperbolas needed to envelope the trajectory enables a classification of orbits to be made. The coalescing of two co-planar ellipses gives rise to an elliptic orbit for which the two attractive masses are the foci. (See Problem 4.1.) A review of the classical theory of the motion of a particle in the presence of two fixed masses has been made by Pauli [41].

4.3 THE GENERALIZED HYDROGEN MOLECULAR ION

Suppose charges Z_Ae and Z_Be are situated at the fixed points A and B. The Hamiltonian for a circulating electron is

$$H = \hbar^2\mathbf{p}^2/2m - Z_Ae^2/r_A - Z_Be^2/r_B.$$

To obtain the Schrödinger equation, we make the replacement $\mathbf{p}^2 \to -\nabla^2$, and use Eq. (4.8) for ∇^2. It is found that on writing

$$\Psi(\mathbf{r}) = \Xi(\xi)\mathrm{N}(\eta)\Phi(\phi) \tag{4.10}$$

for the eigenfunction Ψ, the equation separates. (N is preferred to the overworked symbol H.) The solution for Φ is

$$\Phi(\phi) = e^{im\phi} \qquad (m = 0, \pm1, \pm2, \ldots) \tag{4.11}$$

and the other two functions satisfy the equations

$$\frac{\partial}{\partial\eta}\left\{(1-\eta^2)\frac{\partial \mathrm{N}}{\partial\eta}\right\}+\left\{f-\tfrac{1}{2}\epsilon\rho^2\eta^2-\rho\eta(Z_A-Z_B)-\frac{m^2}{1-\eta^2}\right\}\mathrm{N}=0, \quad (4.12)$$

$$\frac{\partial}{\partial\xi}\left\{(\xi^2-1)\frac{\partial \Xi}{\partial\xi}\right\}+\left\{-f+\tfrac{1}{2}\epsilon\rho^2\xi^2+\rho\xi(Z_A+Z_B)-\frac{m^2}{\xi^2-1}\right\}\Xi=0, \quad (4.13)$$

where the dimensionless quantities ϵ and ρ are related to the eigenvalue E and the distance $2b$ between A and B by the equations

$$E=\epsilon e^2/a_0, \qquad 2b=\rho a_0. \qquad (4.14)$$

The quantity f in Eqs. (4.12)–(4.13) is a separation constant.

The first point to notice about Eqs. (4.12)–(4.13) is that they are of precisely the same form. Thus, if we make the replacements

$$\eta\to\xi, \qquad \mathrm{N}\to\Xi, \qquad Z_A-Z_B\to Z_A+Z_B$$

in the first equation, and then make an overall sign change, we obtain the second one. This similarity goes largely unnoticed for H_2^+, since the term linear in η vanishes when we set $Z_A = Z_B$. The solutions for $\mathrm{N}(\eta)$ in this special case have been studied in detail by Flammer [51], among others. The situation is more complex when the linear term is present, and one usually resorts to a power series expansion, such as that described by Jaffé [52].

Let us defer the question of the detailed solutions for N and Ξ, and turn our attention to the separation constant f. As Erikson and Hill have made clear [53], any exact separation of a Schrödinger equation entails the existence of operators that commute with the Hamiltonian and are therefore constants of the motion. For example, m is the eigenvalue of $-i\,\partial/\partial\phi$, and this commutes with H. To find the operator for which f is an eigenvalue, we multiply Eqs. (4.12) and (4.13) by $\xi^2\Xi$ and $\eta^2\mathrm{N}$ respectively and add, thus eliminating ϵ. On dividing by $\xi^2-\eta^2$, an equation of the form

$$F\Xi\mathrm{N}=f\Xi\mathrm{N}$$

is obtained. The differential operator F is quite complicated, and some simplification must be attempted. Now, for $b\to 0$, we find $\rho\to 0$ and $\eta\to\cos\theta$; furthermore, Eq. (4.12) is satisfied, in this limit, by

$$\mathrm{N}=P_l^m(\eta), \qquad f=l(l+1).$$

This suggests that $\boldsymbol{l}^2$ should be subtracted from F. Using Eqs. (4.9) to construct $\boldsymbol{l}^2$, we find that the part of $F-\boldsymbol{l}^2$ that involves differentiation with respect to ξ and η turns out to be identical to $-b^2\,\partial^2/\partial z^2$. All in all, we get

$$F = l^2 + b^2 p_z^2 + Z_A \rho z / r_A - Z_B \rho z / r_B \tag{4.15}$$

for the operator whose eigenvalue is f and which necessarily commutes with H. This form for F, which has been given by Coulson and Joseph [54], is consistent with Erikson and Hill's rather more cumbersome expression.

4.4 THE HYDROGEN ATOM IN SPHEROIDAL COORDINATES

In spite of their complexity, Eqs. (4.12)–(4.13) might be expected to possess reasonably tractable solutions when $Z_B = 0$, for in this case the equations correspond to a hydrogen atom whose nucleus is at A. This rather ungainly way of looking at a familiar problem is worth pursuing for several reasons. In the first place, the special symmetries of the hydrogen atom may lead us to uncover unexpected structure in the equations. Secondly, it turns out to be possible to construct a certain class of solutions for the generalized hydrogen molecular ion from the spheroidal solutions for the hydrogen atom, as will be discussed in Section 4.7. Furthermore, it would be of considerable interest to find the spectrum of the separation constant f for a simple system such as the hydrogen atom.

We therefore set $Z_B = 0$ in Eqs. (4.12)–(4.13). It is also convenient to write $Z_A = Z$ and $\epsilon = -Z^2/2n^2$, where n is the principal quantum number. The standard approach to equations like (4.12) and (4.13) is to first seek an asymptotic solution. In the limit $\xi \to \infty$, we quickly find $\Xi \to e^{-q\xi}$, where

$$q = \rho(-\epsilon/2)^{1/2} = Z\rho/2n. \tag{4.16}$$

The asymptotic solution is now removed from Ξ by introducing a new function $\Lambda(\xi)$ for which $\Xi = e^{-q\xi}\Lambda$. In an analogous way, it is appropriate to extract $e^{-q\eta}$ from $\mathrm{N}(\eta)$ by writing $\mathrm{N} = e^{-q\eta}\Theta(\eta)$, for then the product $\Xi\mathrm{N}$ yields up the factor

$$\exp(-q(\xi + \eta)) = \exp(-qr_A/b) = \exp(-Zr_A/na_0), \tag{4.17}$$

which we expect for hydrogenic solutions (see Eq. (3.32)).

We have now to consider the expansions of $\Lambda(\xi)$ and $\Theta(\eta)$. The usual criteria for the form of such expansions are rapid convergence and simplicity in the recursion relations satisfied by the coefficients in the expansion. Another criterion is that the terms in the expansion should have a well-defined significance with respect to various groups—or, in short, the terms should be tensors. Given the nature of the present analysis, the last criterion carries more weight than considerations of convergence. As Coulson [55] noticed, the associated Legendre polynomials $P_j^m(\eta)$ (defined in Eq. (1.5)) are an obvious choice for the terms in the expansion of $\Theta(\eta)$. The part $(1 - \eta^2)^{m/2}$ in their definition is of the right form to contribute to the part $\sin^m \theta$ of $Y_{lm}(\theta, \phi)$, as Eqs. (4.4) make clear. In addition, it would be quite feasible to construct a framework in which they appear as tensors. We therefore write

$$N(\eta) = e^{-q\eta}\Theta(\eta) = e^{-q\eta} \sum_{j \geq |m|}^{n-1} c_j P_j^m(\eta), \tag{4.18}$$

and substitute into Eq. (4.12). The upper limit to the sum is fixed by noting the maximum power of η needed to be consistent with Eq. (3.32). The coefficients c_j (for $j \geq |m|$) are found to obey the recurrence relation

$$c_{j+1} \frac{(j+n+1)(j+m+1)}{(2j+3)} - c_j \frac{f + q^2 - j(j+1)}{2q}$$
$$- c_{j-1} \frac{(n-j)(m-j)}{(2j-1)} = 0. \tag{4.19}$$

Although as many as three terms are present, we could scarcely have expected anything better. No manipulation of the general equations (4.12)–(4.13) for the hydrogen molecular ion has yet led to anything simpler than a three-term recursion relation in the final power series expansion. In fact, Eq. (4.19) is of a surprisingly elegant form, being invariant under the interchange $n \leftrightarrow m$.

The secular matrix whose elements are the coefficients in Eq. (4.19) can be symmetrized by writing $c_j = Q_j g_j$, where

$$Q_j = (-1)^j \sqrt{\frac{(2j+1)(j-m)!}{(j+m)!(j+n)!(n-j-1)!}}. \tag{4.20}$$

The new coefficients g_j satisfy the equation

$$g_j \frac{j(j+1) - q^2 - f}{2q} = g_{j+1} \sqrt{\frac{[(j+1)^2 - m^2][n^2 - (j+1)^2]}{(2j+1)(2j+3)}}$$
$$+ g_{j-1} \sqrt{\frac{[j^2 - m^2][n^2 - j^2]}{(2j-1)(2j+1)}}. \tag{4.21}$$

Very surprisingly, the coefficients of the g's on the right are the same as those appearing in the expansion of $A_z Y_{njm}$, as can be seen from Eq. (2.25). This means that the problem in hand must be equivalent to solving the equation

$$(\mathbf{l}^2 - q^2 + 2qA_z)\psi = f\psi \tag{4.22}$$

for an eigenvalue f by expanding ψ in four-dimensional spherical harmonics $Y_{njm}(\Omega)$.

It is not difficult to work out what the detailed connection must be. The differential equation for $\Theta(\eta)$ can be obtained from Eq. (4.22) by first setting

$$\psi = \sum_j g_j Y_{njm}(\chi\theta\phi).$$

Next, Eqs. (1.6) and (2.16) are used to replace $\mathbf{l}^2$ and A_z by differential operators; differentiation with respect to χ and ϕ is then carried out. Finally, we make the substitutions

$$\chi = i\infty, \qquad \phi = 0, \qquad \cos\theta = \eta,$$

and the required differential equation is obtained. At the same time,

$$Y_{njm}(i\infty, \cos^{-1}\eta, 0) = w_{nm}Q_jP_j^m(\eta),$$

where w_{nm} is independent of j. It is convenient to write

$$\mathcal{P}_j^m(\eta) = Q_jP_j^m(\eta), \tag{4.23}$$

and the solution for $\mathrm{N}(\eta)$ takes the form

$$\mathrm{N}_{nfm}(\eta) = e^{-q\eta} \sum_j g_j\mathcal{P}_j^m(\eta) \tag{4.24}$$

for specific quantum numbers n, f, and m.

A precisely similar analysis can be carried out for $\Xi(\xi)$. The only new point concerns the definition of $P_j^m(\xi)$, since $\xi \geq 1$. Following Smythe [56], we amend Eq. (1.5) to

$$P_l^m(\mu) = \frac{(\mu^2-1)^{m/2}}{2^l l!}\frac{d^{l+m}}{d\mu^{l+m}}(\mu^2-1)^l \qquad (|\mu| \geq 1). \tag{4.25}$$

Just as before, we define $\mathcal{P}_j^m(\xi)$ as $Q_jP_j^m(\xi)$, and we arrive at the result

$$\Xi_{nfm}(\xi) = e^{-q\xi} \sum_{j\geq|m|}^{n-1} g_j\mathcal{P}_j^m(\xi), \tag{4.26}$$

in which the coefficients g_j are identical to those in Eq. (4.24). The (unnormalized) eigenfunctions Ψ for the hydrogen atom are thus given by

$$\Psi_{nfm}(\xi\eta\phi) = e^{im\phi}\Xi_{nfm}(\xi)\mathrm{N}_{nfm}(\eta). \tag{4.27}$$

Solutions for Ψ, in which Ξ and N are described by power series in $(\xi - 1)$

and $(1 - \eta)$ respectively, have been discussed by Coulson and Robinson [57, 58].

As an example of Eq. (4.27), we consider the three degenerate hydrogenic levels for which $n = 3$ and $m = 0$. For this case, the secular determinant for f, as derived from Eq. (4.21), takes the form

$$\begin{vmatrix} u & (8/3)^{1/2} & 0 \\ (8/3)^{1/2} & u - 2v & (4/3)^{1/2} \\ 0 & (4/3)^{1/2} & u - 6v \end{vmatrix} = 0, \tag{4.28}$$

where $u = (q^2 + f)/2q$ and $v = 1/2q$. Suppose we pick $Z = 6$ and $\rho = (10)^{1/2}/3$, so that $v = 3(40)^{-1/2}$. (This particular choice has relevance for the generalized H_2^+ ion, as will be seen in Section 4.7.) The cubic for u has the roots

$$u = (5/2)^{1/2}, \quad 7(40)^{-1/2} \pm (241/40)^{1/2}.$$

For the first root, $f = 20/9$; and the (unnormalized) eigenfunction for $N(\eta)$ is

$$\begin{aligned} N(\eta) &= \exp(-(10)^{1/2}\eta/3)\ (4(2)^{1/2}\mathcal{P}_0^0 - (30)^{1/2}\mathcal{P}_1^0 - 5\mathcal{P}_2^0) \\ &= (3/32)^{1/2} \exp(-(10)^{1/2}\eta/3)\ (7 + 2\eta(10)^{1/2} - 5\eta^2). \end{aligned} \tag{4.29}$$

Thus a complete (unnormalized) hydrogenic eigenfunction corresponding to $(nfm) = (3, 20/9, 0)$ is

$$\Psi = \exp(-(10)^{1/2}(\xi + \eta)/3)\ (7 + 2\eta(10)^{1/2} - 5\eta^2)(7 + 2\xi(10)^{1/2} - 5\xi^2). \tag{4.30}$$

4.5 THE SPECTRUM OF THE SEPARATION CONSTANT f

The two commuting vectors $\mathbf{K}_1$ and $\mathbf{K}_2$, defined in Eqs. (2.3), can be used to put the eigenvalue equation (4.22) in a more suggestive form. We get

$$l^2 - q^2 + 2qA_z = (\mathbf{K}_1 + \mathbf{K}_2)^2 - q^2 + 2q(K_{1z} - K_{2z}),$$

and the spherical harmonics Y, in terms of which the eigenvectors ψ are expanded, exhibit the same coupling as the kets $|(K_1K_2)jm\rangle$. From a purely mathematical point of view, the problem is the same as an atomic LS multiplet subjected to a magnetic field $\mathbf{H}$ along the z axis. The Hamiltonian H' in this case is

$$H' = \lambda\mathbf{S}\cdot\mathbf{L} + \beta H_z(L_z + 2S_z),$$

where β is the Bohr magneton and λ is a spin-orbit constant. Apart from constant diagonal terms, H' is the same as

$$\tfrac{1}{2}\lambda(\mathbf{S} + \mathbf{L})^2 + \tfrac{1}{2}\beta H_z(S_z - L_z),$$

and we can make the identifications $\mathbf{K}_1 \equiv \mathbf{S}$, $\mathbf{K}_2 \equiv \mathbf{L}$, and $\beta/\lambda = 2q$. Another equivalent problem is the effect of a magnetic field on a hyperfine multiplet. It is well known in atomic physics that no simple solutions exist to these problems, and one usually has recourse to computers to find the eigenvalues of H'. This means that the spectrum of f is not elementary, and, what is more significant, its spectrum for the hydrogen molecular ion can be expected to be even more remote from a simple form. This is not to say that f cannot be calculated for H_2^+, of course; indeed, numerical values of $f + q^2$ have been tabulated by Madsen and Peek [47] for 20 low-lying levels of H_2^+. But it seems unlikely that these results can be cast in an analytic form.

We have not yet considered why Eq. (4.22) possesses the strikingly simple form that it does. To understand this, we return to the operator F, given in Eq. (4.15), and set $Z_B = 0$. This corresponds to a hydrogen atom centered at A. If we introduce the angular momentum $\boldsymbol{l}_A$ of the electron about A, and use the transformation (4.3), we get

$$\begin{aligned} \boldsymbol{l}^2 &= \boldsymbol{l}_A^2 + b[(\boldsymbol{l}_A \times \mathbf{p})_z - (\mathbf{p} \times \boldsymbol{l}_A)_z] + b^2(p_x^2 + p_y^2) \\ &= \boldsymbol{l}_A^2 + 2bp_0a_z - 2Z_Az'b/r_Aa_0 + b^2(p_x^2 + p_y^2), \end{aligned}$$

where the Lenz vector $\mathbf{a}$ is defined in Eq. (3.37). We remove $\mathbf{p}^2$ by introducing the Hamiltonian H_{H} of the hydrogen atom, thereby obtaining

$$F_{\mathrm{H}} = \boldsymbol{l}_A^2 + 2bp_0a_z + (2mb^2/\hbar^2)H_{\mathrm{H}}, \tag{4.31}$$

a result due to Coulson and Joseph [59]. (A quantity subscripted by H refers to the special case $Z_B = 0$, corresponding to a hydrogen atom.) The eigenvalue equation

$$F_{\mathrm{H}}\Psi = f\Psi$$

is now equivalent to Eq. (4.22), since we can express $\boldsymbol{l}_A$ and a_z in terms of the two commuting vectors $\mathbf{k}_1$ and $\mathbf{k}_2$ with the aid of Eqs. (3.38), and the two eigenvalue equations for f become structurally identical. The coefficients match perfectly, for

$$2q = Z\rho/n = 2bp_0,$$

and

$$-q^2 = -Z^2b^2/n^2a_0^2 \equiv (2mb^2/\hbar^2)H_{\mathrm{H}}$$

for a given level n. So either equation gives the same spectrum for f—which must, of course, be the case.

There is an interesting consequence. The expansion of the eigenfunction Ψ as a linear combination of the ordinary spherical-polar solutions for hydrogen, i.e., the expansion

$$\Psi = \sum_l g_l u_{nlm}(\mathbf{r}), \tag{4.32}$$

must involve the same coefficients g as appeared earlier. For example, the solution (4.30), when converted to a linear combination of the functions $u_{3l0}(\mathbf{r})$, becomes, when normalized,

$$(87)^{-1/2}[4(2)^{1/2}u_{3s} - (30)^{1/2}u_{3p} - 5u_{3d}],$$

and the coefficients in the brackets are identical to those of the $\mathcal{P}_n{}^0$ in Eq. (4.29). (See also Problem 4.2.)

4.6 A TWO-CENTER SYSTEM WITH $Z_A \gg Z_B$

Although the previous analysis has been concerned solely with the hydrogen atom, it contains within it the exact nonrelativistic solutions to the two-center system in the limit when $Z_A/Z_B \to \infty$. To see this, we note first that

$$F = F_{\mathrm{H}} - Z_B\rho z/r_B, \qquad H = H_{\mathrm{H}} - Z_B e^2/r_B.$$

Since $[F, H] = 0$, we must have

$$\langle nfm \mid [F_{\mathrm{H}} - Z_B\rho z/r_B, H_{\mathrm{H}} - Z_B e^2/r_B] \mid n'f'm' \rangle = 0, \tag{4.33}$$

where the bra and the ket correspond to the hydrogenic solutions (4.27). Suppose $n' = n$. Then H_{H} produces the same eigenvalue when acting on bra or ket, and it can be replaced by a number. We also note that all position coordinates commute with one another, and this enables us to simplify Eq. (4.33) to

$$\langle nfm \mid F_{\mathrm{H}}(Z_B e^2/r_B) - (Z_B e^2/r_B)F_{\mathrm{H}} \mid nf'm' \rangle$$
$$= (f - f')\langle nfm \mid (Z_B e^2/r_B) \mid nf'm' \rangle = 0.$$

We can deduce that matrix elements of r_B^{-1} off-diagonal in f (but diagonal in n) vanish. They must always vanish when $m \neq m'$, as can be immediately seen by integrating over ϕ. It follows, then, that the matrix of $Z_B e^2/r_B$ is diagonal within the limited basis provided by the collection of eigenfunctions Ψ_{nfm} with definite n.

If Z_B is sufficiently small compared to Z_A, then a natural way to approach this two-center problem is to solve for the case of $Z_B = 0$ and then add $-Z_Be^2/r_B$ as a small perturbation. To first order, no matrix elements off-diagonal in n are required. This fact, together with the result just obtained concerning f and m, means that the hydrogenic eigenfunctions Ψ_{nfm} are solutions to the two-center problem when $Z_A \gg Z_B$.

An interesting situation arises as Z_B moves to infinity, since this corresponds to a hydrogen atom in a constant weak electric field. If we set $q \to \infty$ in Eq. (4.22), the term in A_z dominates over the term in $\mathbf{l}^2$, and

$$\psi \to | K_1m_1, K_2m_2\rangle,$$

where m_1 and m_2 are the eigenvalues of K_{1z} and K_{2z}. Thus the coefficients g_j become the CG coefficients

$$(K_1m_1, K_2m_2 \mid jm),$$

in which $K_1 = K_2 = k = \frac{1}{2}(n-1)$ [60]. For a given m, the various differences $m_1 - m_2$ are the eigenvalues of A_z, and they serve to distinguish the different solutions. They are more convenient to use than f, since

$$m_1 - m_2 = \lim_{q\to\infty}(f + q^2)/2q.$$

The usefulness of labeling the Stark states of a hydrogen atom by the eigenvalues of A_z was known to Pauli [40].

The eigenfunctions of a hydrogen atom in a weak electric field take an interesting form in momentum space. From Eqs. (2.38) and (3.31),

$$\begin{aligned}\sum_l (km_1, km_2 \mid lm)\psi_{nlm}(\mathbf{p}) &= 4p_0^{5/2}(p^2 + p_0^2)^{-2} \sum_l (km_1, km_2 \mid lm) Y_{nlm}(\Omega) \\ &= 4p_0^{5/2}(p^2 + p_0^2)^{-2}(2\pi^2)^{-1/2}D_{m_1m_2}{}^{(kk)}(\omega),\end{aligned}$$

where the Euler angles ω are related to Ω through the two sets of equations (2.14) and (3.24). Thus the $\mathfrak{D}$ functions reappear in a new role (see Englefield [61]).

4.7 ELEMENTARY SOLUTIONS TO THE GENERALIZED HYDROGEN MOLECULAR ION

The two parts N and Ξ of the eigenfunction for the generalized hydrogen molecular ion satisfy two distinct equations, namely Eqs. (4.12)–(4.13). They are connected only in having ϵ and f in common, since Z_A and Z_B appear as two independent linear combinations. We have already examined the relatively simple eigenfunctions N and Ξ that arise in the hydrogenic case corresponding to $Z_B = 0$. However, it was noticed by Coulson and

Robinson [57] that the two solutions for N and Ξ could both be hydrogenic without implying that $Z_B = 0$. For two solutions N and Ξ characterized by principal quantum numbers n_N and n_Ξ, we must evidently have

$$\epsilon = -\frac{(Z_A - Z_B)^2}{2n_N^2} = -\frac{(Z_A + Z_B)^2}{2n_\Xi^2}. \tag{4.34}$$

The condition that f be the same for both equations imposes a condition on ρ, the separation between the centers. For $n_N \neq n_\Xi$, we must necessarily have $Z_B \neq 0$, and so we have constructed an exact solution to the two-center problem. Solutions of this kind have been called *elementary* by Demkov [62], who has studied their properties in detail.

As an example, consider the $m = 0$ solution for $n_N = 2$ and $n_\Xi = 3$. From Eq. (4.34), the simplest integral solution for the charges is $Z_A = 5$ and $Z_B = 1$. The Ξ equation corresponds to a hydrogenic solution for which $n = 3$ and $Z = 6$, and the determinantal equation for f is identical to Eq. (4.24). The equation for N corresponds to $n = 2$ and $Z = 4$; in this case the determinantal equation for f is

$$\begin{vmatrix} u & 1 \\ 1 & u - 2v \end{vmatrix} = 0. \tag{4.35}$$

The two equations (4.28) and (4.35) are sufficient to solve for f and also for ρ. When this is done, we find $f = 20/9$ and $\rho = (10)^{1/2}/3$. In preparation for this result, we have already constructed in Section 4.4 the solution $\Xi(\xi)$ corresponding to these parameters. The new solution for $N(\eta)$ is

$$N_{nfm}(\eta) = \exp(-(10)^{1/2}\eta/3)\,[5\eta + (10)^{1/2}]$$

for $(nfm) \equiv (2, 20/9, 0)$. Thus an exact (unnormalized) eigenfunction for an electron circulating around the two charges $6e$ and e, fixed a distance $a_0(10)^{1/2}/3$ apart, is

$$\exp(-(10)^{1/2}(\xi + \eta)/3)\,[5\eta + (10)^{1/2}][5\xi^2 - 2(10)^{1/2}\xi - 7]. \tag{4.36}$$

The corresponding eigenvalue E is $-2e^2/a_0$. Although the elementary solutions represent only a very small fraction of all possible solutions, they have a considerable intrinsic interest, as well as being of great potential value in checking numerical calculations.

4.8 THE MANY-CENTER PROBLEM

A particularly elegant analysis has been made by Shibuya and Wulfman [37, 63] of the problem of a single electron in the presence of many fixed nuclei. The central idea is to work in momentum space and expand the electron's eigenfunction as a linear combination of four-dimensional spherical harmonics. Unlike the more familiar position-space analysis, this treatment allows us to apply standard angular momentum techniques to the whole problem, and not merely to the analysis of the angular variables.

If charges $Z_j e$ are situated at the points $\mathbf{R}_j$, Eq. (3.30) generalizes at once to

$$\varphi(\Omega) = \frac{1}{a_0 p_0} \sum_j \frac{Z_j}{2\pi^2} \int \frac{\exp(-i\mathbf{R}_j \cdot (\mathbf{p} - \mathbf{p}'))}{4 \sin^2 \kappa/2} \varphi(\Omega')\, d\Omega'. \tag{4.37}$$

In expanding the kernel, we use Eq. (2.44) and write

$$Z_j \exp(-i\mathbf{R}_j \cdot (\mathbf{p} - \mathbf{p}'))\ (8\pi^2 \sin^2 \kappa/2)^{-1}$$

$$= Z_j \sum_t [\exp(-i\mathbf{p} \cdot \mathbf{R}_j)\ Y_t(\Omega)] \frac{1}{n} [\exp(-i\mathbf{p}' \cdot \mathbf{R}_j)\ Y_t(\Omega')]^*, \tag{4.38}$$

in which the abbreviation $t \equiv (nlm)$ has been made. The expansion

$$\exp(i\mathbf{p} \cdot \mathbf{R})\ Y_t(\Omega) = \sum_\tau S_\tau{}^t(\mathbf{R}) Y_\tau(\Omega), \tag{4.39}$$

where $\tau \equiv (\nu\lambda\mu)$ is now introduced. In order to outline Shibuya and Wulfman's method with as few digressions as possible, we postpone for the moment the evaluation of the quantities $S_\tau{}^t(\mathbf{R})$. However, we see that

$$S_\tau{}^t(\mathbf{R}) \doteq \int \exp(i\mathbf{p} \cdot \mathbf{R})\ Y_t(\Omega) Y_\tau{}^*(\Omega)\, d\Omega, \tag{4.40}$$

and so

$$S_\tau{}^t(\mathbf{R}) = [S_t{}^\tau(-\mathbf{R})]^*. \tag{4.41}$$

If we replace $\mathbf{R}$ by $-\mathbf{R}$ in Eq. (4.39), we get

$$\exp(-i\mathbf{p} \cdot \mathbf{R})\ Y_t(\Omega) = \sum_\tau [S_t{}^\tau(\mathbf{R})]^* Y_\tau(\Omega),$$

and so the right-hand side of Eq. (4.38) becomes

$$Z_j \sum_{t\tau\tau'} [S_t{}^\tau(\mathbf{R}_j)]^* \frac{1}{n} S_t{}^{\tau'}(\mathbf{R}_j) Y_\tau(\Omega) Y_{\tau'}{}^*(\Omega'). \tag{4.42}$$

The next step is to express the solution $\varphi(\Omega)$ that we are seeking as a linear combination of the four-dimensional harmonics:

$$\varphi(\Omega) = \sum_a c_a Y_a(\Omega).$$

When the kernel (4.42) is put back into Eq. (4.37), the integration over

Ω' can be performed. Selecting the coefficient of $Y_b(\Omega)$ on both sides of the integral equation, we get

$$c_b = (a_0 p_0)^{-1} \sum_j Z_j \sum_{at} [S_t^b(\mathbf{R}_j)]^* \frac{1}{n} S_t^a(\mathbf{R}_j) c_a, \tag{4.43}$$

that is,

$$p_0 c_b = a_0^{-1} \sum_a P_b^a c_a, \tag{4.44}$$

where

$$P_b^a = \sum_j Z_j \sum_t [S_t^b(\mathbf{R}_j)]^* \frac{1}{n} S_t^a(\mathbf{R}_j). \tag{4.45}$$

Once the $S(\mathbf{R}_j)$ are known, the quantities P_b^a can be found. We then set up the secular equations (4.44) from which the eigenvalue p_0 (representing the energy through the relation $E = -\hbar^2 p_0^2/2m$) and the coefficients c_a (through which the eigenfunction is defined) can be determined. Unlike the solution for a hydrogen atom, the matrix P (whose elements are P_b^a) will not, in general, be diagonal. To obtain a manageable problem, some truncation is necessary. An obvious choice is to limit a and b to those quantum numbers (nlm) for which n is less than some specified figure.

4.9 EVALUATION OF $S_\tau^t(\mathbf{R})$

There is only one integral involving $\exp(i\mathbf{p}\cdot\mathbf{R})$ that we can easily do, namely

$$(2\pi)^{-3/2} \int \exp(i\mathbf{p}\cdot\mathbf{R})\, \psi_T(\mathbf{p})\, d\mathbf{p} = u_T(\mathbf{R}). \tag{4.46}$$

This is just a statement that the Fourier transform of the momentum-space eigenfunction ψ for a hydrogenic solution T ($\equiv NLM$) is the corresponding position-space eigenfunction u. There is a certain amount of leeway in using this equation, since the nuclear charge $Z'e$ characterizing ψ and u need not be an integral multiple of e. It is convenient to define

$$p_T = Z'/a_0 N \tag{4.47}$$

for the corresponding value of p_0.

In order to see what is involved in doing the integral of Eq. (4.46), we write the left-hand side of this equation as

$$2(p_T/2\pi)^{3/2} \int \exp(i\mathbf{p}\cdot\mathbf{R})\, [(p^2 + p_T^2)/4p_T^2] Y_T(\Omega)\, d\Omega \tag{4.48}$$

with the aid of Eq. (3.31). We note that

$$(p^2 + p_T{}^2)/4p_T{}^2 = \tfrac{1}{4}(1 + \tan^2 \chi/2),$$

which is a function of Ω only. To convert $S_\tau{}^t(\mathbf{R})$ to the form (4.48), we write

$$Y_t(\Omega)\, Y_\tau{}^*(\Omega) = \sum_T [(p^2 + p_0{}^2)/4p_0{}^2] Y_T(\Omega)\, V_{T\tau}{}^t,$$

where

$$V_{T\tau}{}^t = \int [4p_0{}^2/(p^2 + p_0{}^2)] Y_t(\Omega)\, Y_\tau{}^*(\Omega)\, Y_T{}^*(\Omega)\; d\Omega. \tag{4.49}$$

Thus

$$S_\tau{}^t(\mathbf{R}) = \sum_T V_{T\tau}{}^t \int \exp(i\mathbf{p}\cdot\mathbf{R})\; [(p^2 + p_0{}^2)/4p_0{}^2] Y_T(\Omega)\; d\Omega.$$

This is still not quite in the right form to match the integral (4.48). The quantity $\mathbf{p}\cdot\mathbf{R}$ in the exponential, when expressed as a function of Ω, introduces p_0 through such substitutions as

$$p_x = p_0 \tan \chi/2 \sin \theta \cos \phi,$$

while the exponential in the integral (4.48) involves p_T instead. Fortunately, we can easily correct this by a radial scaling. We write

$$\mathbf{R} = (p_T/p_0)\mathbf{R}', \tag{4.50}$$

and the integral can be performed. We get

$$S_\tau{}^t(\mathbf{R}) = \tfrac{1}{2} \sum_T (2\pi/p_T)^{3/2} V_{T\tau}{}^t u_T(\mathbf{R}'). \tag{4.51}$$

The functions u_T are given explicitly in Eq. (3.32). The argument w of the Laguerre polynomial appearing in that equation is given by

$$w/2 = (Z'R'/Na_0) = p_T R' = p_0 R = s, \tag{4.52}$$

say. This parameter s is a convenient quantity to work with.

The integral $V_{T\tau}{}^t$ can be evaluated without difficulty. We see that

$$\begin{aligned} 4p_0{}^2/(p^2 + p_0{}^2) &= 4/(1 + \tan^2 \chi/2) = 2(1 + \cos \chi) \\ &= (2\pi^2)^{1/2}[2Y_{100}(\Omega) + Y_{200}(\Omega)], \end{aligned}$$

and the product of any pair of harmonics Y_t and $Y_{t'}$ can be combined either by using the general formula (2.33) or by working the product out explicitly. The second method is particularly suitable for the simpler harmonics, as tables of them are available [37].

As examples of a few S functions, we cite the following:

$$S_{200}{}^{200} = \tfrac{1}{3}e^{-s}(3 + 3s - 2s^2 + s^3),$$

$$S_{210}{}^{200} = \tfrac{1}{3}e^{-s}s(1 + s - s^2) \cos\Theta,$$

$$S_{210}{}^{210} = \tfrac{1}{9}e^{-s}[9 + 9s + 2s^2 - s^3 - (s^2 + s^3)(3\cos^2\Theta - 1)].$$

These correspond to a vector $\mathbf{R}$ possessing polar coordinates $(R\Theta\Phi)$ for which $s = p_0 R$. Because of a different phase convention, our functions $S_\tau{}^t$ (which are always real) are related to the functions $S_{\nu\lambda\mu}{}^{nlm}$ of Shibuya and Wulfman [37] through the equation

$$S_\tau{}^t = i^{\lambda - l} S_{\nu\lambda\mu}{}^{nlm}.$$

A useful check on the working is the condition

$$S_\tau{}^t(0) = \delta(t, \tau),$$

which follows at once from the definition (4.40).

The fact that p_0 enters in the relation between s and R is a little awkward. We obviously cannot know p_0 before a particular physical system has been investigated. The procedure to follow is, first, to regard s as a variable parameter and to pick a value for it. The functions $S_\tau{}^t(\mathbf{R}_j)$ can now be computed, taking the angles Θ_j and Φ_j directly from the geometrical arrangement of the fixed charges j. The S functions are fed into Eq. (4.45) to calculate $P_b{}^a$. The range of a and b that one chooses to work with depends on one's persistence and computing facilities. The $P_b{}^a$ enter as elements in the matrix P whose diagonalization yields p_0 and the coefficients c_a of the eigenfunction. Having now found p_0, we can evaluate R from the relation $R = s/p_0$, and thus find the size of the structure that we have been analyzing.

4.10 LIMITED BASES

The crucial factor limiting the accuracy of the values of p_0 is the truncation of the harmonic bases Y_t. To illustrate this, we briefly describe, first, the situation for $H_2{}^+$ when the separation $2b$ between the nuclei is small compared to a_0. This is an interesting case to consider, because the limiting situation $2b/a_0 \to 0$ can be analyzed exactly by standard techniques. The procedure to follow is to imagine the electron of the ion He^+ subjected to the small perturbation

$$H' = -(e^2/r_A) - (e^2/r_B) + (2e^2/r),$$

which converts He^+ into $H_2{}^+$. The radial distances r_A and r_B can be expanded in terms of r and b by using the generating functions for Legendre polynomials (see Eq. 2.42). A straightforward calculation, based on the eigenfunctions $u(\mathbf{r})$ given in Eq. (3.32), yields the energy displacements $(\Delta E)_{nlm}$ produced by H'. For the components of the $n = 2$ levels, they are given, to order $(b/a_0)^2$, by the equations

$$\begin{aligned}(\Delta E)_{200} &= (4/3)b^2(e^2/a_0^3),\\ (\Delta E)_{210} &= -(4/15)b^2(e^2/a_0^3),\\ (\Delta E)_{21\pm1} &= (2/15)b^2(e^2/a_0^3).\end{aligned} \tag{4.53}$$

Let us see how closely these figures can be approached by a momentum-space basis for which $n < 4$. In the limit $b = 0$, the matrix P is diagonal. To the order $(b/a_0)^2$, we need not consider off-diagonal elements, and we have thus to calculate $P_{200}{}^{200}$, $P_{210}{}^{210}$, and $P_{21\pm1}{}^{21\pm1}$. From Eq. (4.45), we see

$$\begin{aligned}P_{200}{}^{200} = {} & |S_{100}{}^{200}|^2 + \tfrac{1}{2}|S_{200}{}^{200}|^2 + \tfrac{1}{2}|S_{210}{}^{200}|^2\\ & + \tfrac{1}{3}|S_{300}{}^{200}|^2 + \tfrac{1}{3}|S_{310}{}^{200}|^2 + \tfrac{1}{3}|S_{320}{}^{200}|^2\end{aligned}$$

for each charge. The limitation $n < 4$ enters here. The S functions are calculated for charges at $(0, 0, \pm b)$, and the values of $P_a{}^a$ give directly the eigenvalues p_0. When these are converted to energy shifts $\Delta E'$, we find

$$\begin{aligned}(\Delta E')_{200} &= (16/9)b^2(e^2/a_0^3),\\ (\Delta E')_{210} &= -(16/75)b^2(e^2/a_0^3),\\ (\Delta E')_{21\pm1} &= (13/75)b^2(e^2/a_0^3).\end{aligned} \tag{4.54}$$

An alternative method for finding these shifts is described in Problem 4.8. There is modest agreement between Eqs. (4.53) and (4.54).

Comparisons of this kind appear in a more favorable light if the actual energy E rather than its displacements is cited. Wulfman [63] has used the more extensive basis corresponding to $n < 6$ to study the energy of the ground state of H_2^+ as a function of the internuclear separation $2b$. The difference between the energy calculated in this way and the essentially exact energy as calculated by Bates and collaborators [44, 45] is found to oscillate about zero as b increases from 0 to just beyond the equilibrium distance (for which $b \sim a_0$). In this region, the discrepancy amounts to at most 0.08 Ry, which is 7% of the energy of the electron for the equilibrium separation.

It should be apparent that the momentum-space analysis cannot compete with the traditional techniques of finding the energies of H_2^+. Its appeal lies in other directions. Most striking is its structural unity, as demonstrated by the use of angular momentum methods to treat both radial and angular aspects of the problem. From a practical point of view, however, it is worth noticing that it is comparatively simple to extend the analysis from H_2^+ to multicenter systems. These cases are not so easily handled by position-space analyses as H_2^+.

4.11 HIGHER GROUPS

The application of group theory to the analysis of a single electron moving in the electrostatic field of many nucleons need not be confined to $R(4)$. Since the eigenfunctions of the electron is expanded in terms of the harmonics Y_{nlm}, the very rich group-theoretical structure of the hydrogen atom can be carried over to the many-nucleon problem. For example, the collection of hydrogenic eigenfunctions for which $n = 1, 2, \ldots, N$ can be taken as basis functions for a single irreducible representation of dimension $N(N + 1)(2N + 1)/6$ of the rotation group $R(5)$ [64, 65]. This group enters in a different way from $R(4)$. The ten generators of $R(5)$ are different for every N, and they do not commute with the Hamiltonian of the hydrogen atom. When we pass to the many-nucleon problem, this distinction is partly lost, as both $R(4)$ and $R(5)$ become noninvariance groups. Wulfman [63] has described how the WE theorem can be applied to $R(5)$, and a number of interesting relations emerge from the analysis. It is possible to go further and introduce groups such as $O(4, 1)$ and $O(4, 2)$. These are noncompact groups which bear some resemblance to the rotation groups $R(5)$ and $R(6)$, the most obvious distinction being that the invariant quadratic form for $O(p, q)$ comprises p positive and q negative terms. Noncompact groups possess unitary irreducible representations of infinite dimension, and both the bound and the continuum states of the hydrogen atom can be classified by them. Wulfman has pointed out that the generators of $O(4, 2)$ enable points on the hyperspheres of different radius p_0 to be connected [63]. As has been mentioned before, much of our knowledge of these groups comes from the work of Barut *et al.* [27].

Striking though the analysis involving groups such as $O(4, 2)$ is, its successes lie more in the development of concepts and in the simplification of calculations than in the production of new numerical results. This is partly because of our present rather inadequate knowledge of the properties of noncompact groups, which makes it imprudent to apply the WE theorem without special analysis. There is also the fact that the irreducible representations of many of the higher groups are so large that there is no opportunity to exploit selection rules between different representations. The comparatively elementary group $O(2, 1)$ has proved to yield some of the most interesting results. Armstrong [28] has shown that the WE theorem can be used for this group, and that the hydrogenic selection rule [66]

$$\langle nlm \mid r^{-s} \mid nl'm \rangle = 0 \qquad (s = 2, 3, \ldots, \mid l - l' \mid + 1)$$

turns out to correspond to the violation of a triangular condition on the analogs of angular momentum quantum numbers.

4.12 THE GROUP $O(2, 1)$ AND THE (m, n) SYMMETRY

The reciprocal character of the magnetic quantum number m and the principal quantum number n revealed itself in Section 2.5, where the combination $A_z Y_{nlm}(\Omega)$ was evaluated. It reappeared in Section 4.4 during the discussion of the two-center problem. To account for the (m, n) symmetry, we must evidently try to assign to n a role similar to that of m. As a first step, we introduce a new angular variable τ and construct the functions

$$e^{in\tau} Y_{nlm}(\chi\theta\phi). \tag{4.55}$$

This at least puts n and m on a par by associating them with the respective angles τ and ϕ. We have now to find an analog (say $\boldsymbol{l}'$) of $\boldsymbol{l}$ whose components l_+', l_-', and l_z', when acting on the functions (4.55), respectively raise n by one unit, lower n by one unit, and leave it unchanged. In fact, we need to parallel the entire analysis of the spherical harmonics $Y_{lm}(\theta\phi)$ given in Section 1.2. The steps to follow are fairly obvious:

(1) We must first decide on the variables to use in constructing $\boldsymbol{l}'$. Since $\boldsymbol{l}$ depends on θ and ϕ, as well as being a linear combination of $\partial/\partial\theta$ and $\partial/\partial\phi$, we choose $\boldsymbol{l}'$ to be a function of χ and τ in which $\partial/\partial\chi$ and $\partial/\partial\tau$ appear linearly.

(2) It is easy to see that

$$l_z' = \frac{1}{i}\frac{\partial}{\partial\tau}, \tag{4.56}$$

for this operator produces the eigenvalue n when acting on the functions (4.55). As such, it is the analog of l_z.

(3) To construct l_+' and l_-', we need to know the derivative of $C_{nl}(\chi)$ with respect to χ. This can be worked out by the usual analytical techniques, or we can convert C_{nl} to $C_{n-l-1}{}^{l+1}$ (see Problem 2.2) and refer to standard texts [67] for the results that we want.

(4) Since $l_\pm'$ must be prefixed by $e^{\pm i\tau}$ to give the right exponential

dependence on τ, it is easy to ensure

$$[l_\pm', l_z'] = \mp l_\pm'.$$

The additional condition that we want to hold, namely

$$[l_+', l_-'] = 2l_z',$$

serves to fix the magnitudes of the shift operators. The result is

$$l_\pm' = e^{\pm i\tau}\left(\sin\chi\frac{\partial}{\partial\chi} \mp i\cos\chi\frac{\partial}{\partial\tau} + \cos\chi\right). \tag{4.57}$$

(5) We require $\mathbf{l}'$ to act on an orthonormal basis. If we extract the part $e^{in\tau}C_{nl}(\chi)$ from (4.55), we must recognize that integration over χ now has a different role to play than it has for the functions $Y_{nlm}(\Omega)$. There, orthogonality between different l functions is guaranteed by the presence of $Y_{lm}(\theta\phi)$, and integration over χ ensures orthogonality between functions of the same l and different n. Here, this last orthogonality is taken care of by integration over τ, and so the integration over χ has to ensure orthogonality between functions of the same n but different l. If we write

$$Z_{ln}(\chi\tau) = (-i)^l\{(2l+1)/4\pi n\}^{1/2}C_{nl}(\chi)e^{in\tau},$$

we find that this basis is orthonormal provided the measure $d\tau\,d\chi$ is used, and the range of integration is given by $0 \le \tau \le 2\pi$, $0 \le \chi \le \pi$. We no longer need the weighted infinitesimal $\sin^2\chi\,d\chi$. (A method of obtaining this result forms the subject of Problem 4.9.) It should be pointed out that the final term $e^{\pm i\tau}\cos\chi$ on the right-hand side of Eq. (4.57) could be removed if the Z functions were multiplied by $\sin\chi$ and the measure $\csc^2\chi\,d\tau\,d\chi$ were used [67a].

(6) The action of $\mathbf{l}'$ on the functions Z is found to be

$$\begin{aligned} l_z'Z_{ln} &= nZ_{ln}, \\ l_\pm'Z_{ln} &= \pm[n(n\pm1) - l(l+1)]^{1/2}Z_{l,n\pm1}. \end{aligned} \tag{4.58}$$

Furthermore,

$$\mathbf{l}'^2 Z_{ln} = l(l+1)Z_{ln}. \tag{4.59}$$

In spite of their familiar appearance, these equations are quite different from the corresponding ones for $R(3)$. The inequality $n > l$ means that the shift operator l_+' can be applied continually to a function Z_{ln}, thereby augmenting n in unit steps, without ever producing a null result. The representation (which is irreducible) is thus of infinite dimension, in contrast to the situation that we are familiar with in $R(3)$.

Another difference springs from the possible negative sign just to the right of the equality sign in the second of the two equations (4.58). It is easy to verify, for example, that l_x' and l_y' are anti-Hermitian within the basis provided by the functions Z_{ln}. This contrasts with the Hermiticity of l_x and l_y taken within the Y_{lm} basis. If we define $\mathbf{l}''$ by writing

$$l_x'' = il_x', \qquad l_y'' = il_y', \qquad l_z'' = l_z',$$

then $\mathbf{l}''$ is Hermitian; but the commutation relations for the Cartesian components of $\mathbf{l}''$ exhibit at least one reversal of sign when compared to the commutation relations of the components of $\mathbf{l}$. This property is characteristic of the noncompact group $O(2, 1)$, since the sign reversals occur naturally when generators are constructed to preserve a point (xyz) on the hyperboloid $x^2 + y^2 - z^2 = 1$ rather than on the sphere $x^2 + y^2 + z^2 = 1$ (see Problem 4.10).

Not every property of the system we are studying differs from that of $R(3)$. We can rapidly convert an element of the group $O(2, 1)$, such as

$$\exp\left[(\Delta\tau)\frac{\partial}{\partial\tau} + (\Delta\chi)\frac{\partial}{\partial\chi}\right],$$

where $\Delta\tau$ and $\Delta\chi$ are two *real* parameters, to the form e^{iU}, where U is Hermitian. The representation provided by the Z functions is therefore unitary, just as that provided by the Y functions is unitary in the case of $R(3)$. However, it is precisely here that a crucial distinction between $O(2, 1)$ and $R(3)$ lies. For though we may define bases and operators in various ways (changing, for example, the form of the commutation relations satisfied by the components of $\mathbf{l}'$), the unitary irreducible representations of $O(2, 1)$ are of infinite dimension, while those of $R(3)$ are always finite.

We are now in a position to examine the (m, n) symmetry exhibited by Eq. (2.25). If we use the Z functions, this equation becomes

$$\begin{aligned} A_z Y_{lm} Z_{ln} = {} & (2l+3)^{-1}[(l+1)^2 - m^2]^{1/2}[n^2 - (l+1)^2]^{1/2} Y_{l+1,m} Z_{l+1,n} \\ & + (2l-1)^{-1}[l^2 - m^2]^{1/2}[n^2 - l^2]^{1/2} Y_{l-1,m} Z_{l-1,n}. \qquad (4.60) \end{aligned}$$

The m dependence of the coefficients stems from the fact that $\mathbf{A}$ is a vector with respect to $\mathbf{l}$. To find how it stands with respect to $\mathbf{l}'$, we first calculate

$$[l_\pm', A_z],$$

which we call $A_\pm'$. It is straightforward to do this, and also to prove that a repetition of the calculation produces zero:

$$[l_+', A_+'] = [l_-', A_-'] = 0.$$

Thus A_+', A_z and A_-' form the components of a vector $\mathbf{A}'$ with respect to $O(2, 1)$. (The possibility that A_z contains a scalar component can be eliminated by showing that

$$[l_-', A_+'] = A_z'.)$$

Thus A_z is the component $T_{00}^{(11)}$ of a double tensor $\mathbf{T}^{(11)}$ having one foot in $R(3)$ and the other in $O(2, 1)$. Armstrong [28] has shown that the WE theorem can be applied to $O(2, 1)$ and that the CG coefficients of this group are algebraically identical to those of $R(3)$. This immediately accounts for the similarity between the two factors making up the coefficients on the right-hand side of Eq. (4.60). It is worth mentioning that $\boldsymbol{l}' \cdot \mathbf{A}' = 0$, a result which parallels $\boldsymbol{l} \cdot \mathbf{A} = 0$. No doubt much of the analysis for $R(4)$ could be repeated for the group whose generators are $\boldsymbol{l}'$ and $\mathbf{A}'$, but this would take us too far afield.

PROBLEMS

4.1 A classical particle of mass m moves in a plane under the gravitational attraction of two fixed masses that are separated by a distance $2b$. Prove that the kinetic energy T is given in terms of spheroidal coordinates by

$$T = \frac{1}{2} mb^2 \left(\frac{\xi^2 - \eta^2}{\xi^2 - 1} \dot{\xi}^2 + \frac{\xi^2 - \eta^2}{1 - \eta^2} \dot{\eta}^2 \right),$$

and construct the Lagrangian L. Prove that Lagrange's equations are consistent with the elliptical solution $\ddot{\xi} = \dot{\xi} = 0$, and show that, for equal fixed masses, the velocity of the particle in its orbit is proportional to

$$\{(\xi^2 + \eta^2)/(\xi^2 - \eta^2)\}^{1/2}.$$

4.2 Examine the properties of the functions $\Xi(\xi)$ and $\mathrm{N}(\eta)$ (given in Eqs. (4.26) and (4.24)) in the limit $\xi \to \infty$. Show that they go over into a function depending on r and one depending on $\cos \theta$, and hence prove that the expansion (4.32) for Ψ follows from the condition that spheroidal coordinates become polar coordinates in the limit of large ξ.

4.3 A two-center system for which $Z_A \gg Z_B$ is characterized by a separation $2b$. Prove that, for very large b, the unnormalized eigenfunction of a circulating electron can be approximated by

$$\exp im\phi \exp(-Zr/na_0) \ (\xi + 1)^{k+t}(\xi - 1)^{k-t}(1 + \eta)^{k+t}(1 - \eta)^{k-t},$$

where $k = \frac{1}{2}(n - 1)$ and $t = m_1 - m_2$, in the notation of Section 4.6. Investigate the impossibility of normalizing the eigenfunction in the form given above. (Parabolic coordinates are the appropriate ones to use in this limiting case [68].)

4.4 An electron circulates around two charges of $3e$ and $2e$, which are fixed a distance $a_0\rho$ apart. Prove that an elementary solution exists corresponding to an eigenfunction

$$\exp(-\tfrac{1}{2}(15)^{1/2}(\xi + \eta))\,[15\xi^4 - 20(15)^{1/2}\xi^3 + 90\xi^2 + 12(15)^{1/2}\xi - 81],$$

an energy $-e^2/2a_0$, and a separation defined by $\rho = (15)^{1/2}$. Prove that the above polynomial in ξ, when expressed as a linear combination of the functions $P_j(\xi)$, contains no term for which $j = 1$. Show that this result is characteristic of all solutions for which $f = \tfrac{1}{2}\rho^2\epsilon$, in the notation of Section 4.3.

4.5 Show that the S functions of Section 4.8 can be written in the form

$$S_t^a(\mathbf{s}) = [t \mid \exp i\mathbf{p}\cdot\mathbf{R} \mid a],$$

in which the square-bracketed ket (and bra) denote a four-dimensional spherical harmonic (and its complex conjugate) for which $a \equiv (nlm)$, etc. If an operator Π is introduced for which

$$[t \mid \Pi \mid t'] = (a_0 n)^{-1}\delta(t, t'),$$

show that Eq. (4.44) can be put in the operator form

$$p_0\varphi = P\varphi,$$

where

$$P = \sum_j Z_j \exp(-i\mathbf{p}\cdot\mathbf{R}_j)\,\Pi \exp(i\mathbf{p}\cdot\mathbf{R}_j).$$

Derive the equations

$$\begin{aligned}\langle nlm \mid \Pi \mid nl'm'\rangle &= (a_0 n)^{-1}\delta(l, l')\delta(m, m')\\ &= \delta(l, l')\delta(m, m')\,\langle nlm \mid p^2 \mid nlm\rangle^{1/2},\end{aligned}$$

where the bras and kets in these equations denote hydrogenic eigenstates (see Wulfman [63]).

4.6 The operators $\Pi^{(\alpha)}$ are defined iteratively by

$$\Pi^{(\alpha+1)} = [\Pi^{(\alpha)}, i\mathbf{p}\cdot\mathbf{R}],$$

with $\Pi^{(0)} = \Pi$ (defined in Problem 4.5). Show that

$$\exp(-i\mathbf{p}\cdot\mathbf{R})\,\Pi \exp i\mathbf{p}\cdot\mathbf{R} = \sum_{\alpha=0}^{\infty} \frac{1}{\alpha!}\,\Pi^{(\alpha)}.$$

Prove

$$[t \mid \Pi^{(1)} \mid t] = 0$$

(see Wulfman [63]).

4.7 By using the generating functions for Gegenbauer polynomials, obtain the expansions

$$\tan \chi/2 \cos \theta = 2\pi i \sum_{n>1} \{2/3(n^2 - 1)\}^{1/2} Y_{n10}(\Omega),$$

$$\tan^2 \chi/2 = -1 - 4\pi(2)^{1/2} \sum_{n} (-1)^n Y_{n00}(\Omega),$$

$$\tan^2 \chi/2\ (3 \cos^2 \theta - 1) = 4\pi \sum_{n>2} (-1)^n \{2(n^2 - 4)/5(n^2 - 1)\}^{1/2} Y_{n20}(\Omega),$$

$$\csc^2 \chi = 2\pi(2)^{1/2} \sum_{n \text{ odd}} Y_{n00}(\Omega).$$

4.8 Evaluate

$$[2lm \mid \Pi^{(2)} \mid 2lm],$$

where $\Pi^{(2)}$ is defined in Problem 4.6, and hence check the energy shifts $(\Delta E')_{nlm}$ given in Section 4.10 for H_2^+ in the limit of small internuclear separation.

4.9 An atomic configuration comprises two equivalent electrons with azimuthal quantum number a. The operator $\mathbf{C}_1^{(l)} \cdot \mathbf{C}_2^{(l)}$ ($= V$, say) splits the configuration into terms, each one characterized by a total orbital angular momentum x. Calculate

$$\sum_x \langle (aa)x \mid V \mid (aa)x \rangle$$

by (i) using Eq. (1.67), and (ii) finding the trace of the matrix of V in the basis characterized by $M_x = 0$. Equate the two results and hence obtain

$$\sum_x (-1)^x \begin{Bmatrix} a & a & l \\ a & a & x \end{Bmatrix} = (-1)^{l+2a} \frac{1}{2l+1}.$$

(The absence of the usual weighting factor $2x + 1$ in the sum is to be noted.) Use this equation and the last expansion of Problem 4.7 to evaluate $\csc^2 \chi Y_{nlm}(\Omega)$. Hence derive

$$\int_0^\pi C_{nl}(\chi) C_{nl'}(\chi)\, d\chi = 2n\delta(l, l')/(2l + 1).$$

4.10 A point (xyz) on the surface of the hyperboloid of revolution

$$x^2 + y^2 - z^2 = 1$$

is parametrized by setting

$$x = \csc \chi \sin \tau, \qquad y = -\csc \chi \cos \tau, \qquad z = \cot \chi.$$

Prove that the infinitesimal displacements of (xyz) produced by the components of $\mathbf{j}$, where

$$j_x = \frac{1}{i}\left(y\frac{\partial}{\partial z} + z\frac{\partial}{\partial y}\right), \qquad j_y = \frac{1}{i}\left(z\frac{\partial}{\partial x} + x\frac{\partial}{\partial z}\right), \qquad j_z = \frac{1}{i}\left(x\frac{\partial}{\partial y} - y\frac{\partial}{\partial x}\right),$$

leave the point on the hyperboloid. Show that the commutation relations

$$[j_x, j_y] = ij_z, \qquad [j_y, j_z] = -ij_x, \qquad [j_z, j_x] = -ij_y$$

are satisfied. Define $j_\pm = ij_x \pm j_y$, and prove that the usual commutation relations

$$[j_z, j_\pm] = \pm j_\pm, \qquad [j_+, j_-] = 2j_z$$

are obtained. Show that

$$j_\pm = e^{\pm i\tau}\left(\sin\chi\frac{\partial}{\partial\chi} \mp i\cos\chi\frac{\partial}{\partial\tau}\right),$$

$$j_z = \frac{1}{i}\frac{\partial}{\partial\tau}.$$

4.11 A vector $\mathbf{R}$ is defined through

$$(R\Theta\Phi) = (s/p_0, 0, 0).$$

Show how the first expansion of Problem 4.7 can be used to evaluate $\partial S_\tau{}^t(R)/\partial s$. Obtain the equation

$$\frac{\partial}{\partial s}[(6)^{1/2}S_{211}{}^{nlm} + S_{311}{}^{nlm}] + (6)^{1/2}S_{321}{}^{nlm} = 0,$$

and generalize it as far as possible.

4.12 Show that $3\cot^2\chi + 1$ transforms like the component of a tensor of rank 2 with respect to the generators $\boldsymbol{l}'$ of $O(2, 1)$ given in Eqs. (4.56)–(4.57). Deduce that

$$[nlm \mid \csc^2\chi(3\cot^2\chi + 1) \mid nlm]$$

for a fixed l and m has a dependence on n given by

$$n\{3n^2 - l(l+1)\},$$

where the square-bracketed bra and ket have the same significance as in Problems 4.5–4.6. Prove $P_k(i\cot\chi)$ is the component of a tensor of rank k, and generalize the above result.

5

Expansions

5.1 GENERAL METHOD

As systems with more than one nucleus or more than one electron are considered, it frequently becomes necessary to convert operators or eigenfunctions to expressions in which the coordinates of the particles appear in a more convenient or accessible form. The addition theorems of Section 2.9 and the expansions of r_{12}^{-1} and r_{12}^{-2} given in Eqs. (2.42)–(2.43) can be thought of as examples of this. There is an extensive literature on the derivation of such expansions, though the methods used are sometimes unnecessarily cumbersome. This is largely because the symmetry properties that these expressions must possess are not always used in a constructive way. It therefore seems well worthwhile to show how this deficiency can be remedied. In what follows, a variety of problems are taken up. A general approach is common to all, and it is probably best illustrated by an example.

For this purpose, we take $\exp i\mathbf{r}_1 \cdot \mathbf{r}_2$, where $\mathbf{r}_1$ and $\mathbf{r}_2$ are vectors of mutually reciprocal dimensionality in ordinary three-dimensional space. The first point to settle is the form for the expansion. Suppose we require the angular dependence on $\mathbf{r}_1$ and $\mathbf{r}_2$ to be represented through the functions $C_m^{(l)}(\theta_1\phi_1)$ and $C_{m'}^{(l')}(\theta_2\phi_2)$ respectively. We have now to determine the functions of the purely radial quantities r_1 and r_2 that must be attached to these harmonics. To do this for $\mathbf{r}_1$, we suppose $\mathbf{r}_2$ to be fixed. Then $\exp(i\mathbf{r}_1\cdot\mathbf{r}_2)$ is a solution ψ_1 to

$$(\nabla_1^2 + r_2^2)\psi_1 = 0.$$

We now write

$$\psi_1 = j(r_1r_2)C_m^{(l)}(\theta_1\phi_1)$$

and find the differential equation that the radial function j must satisfy. By writing r_1r_2 for the argument of j instead of just r_1, the equation simplifies to

$$\left[\frac{1}{x^2}\frac{d}{dx}\left(x^2\frac{d}{dx}\right) + 1 - \frac{l(l+1)}{x^2}\right]j(x) = 0, \tag{5.1}$$

where $x = r_1r_2$. We may now identify $j(x)$ with the spherical Bessel functions $j_l(x)$ or $j_{-l-1}(x)$ (or with any combination of the two). If Eq. (5.1) is not considered recognizable, we can write

$$j_l(x) = (\pi/2x)^{1/2}J_{l+1/2}(x),$$

and the functions $J_\nu(x)$ are found to satisfy Bessel's equation

$$\left[\frac{d^2}{dx^2} + \frac{1}{x}\frac{d}{dx} + 1 - \left(\frac{\nu}{x}\right)^2\right]J_\nu(x) = 0.$$

In order to have a finite expression when $r_1 = 0$, we select only those functions $j_l(x)$ with $l \geq 0$. Thus

$$\psi_1 = j_l(r_1r_2)C_m^{(l)}(\theta_1\phi_1).$$

For r_2, entirely similar functions ψ_2 are obtained. The next step is to form the appropriate products $\psi_1\psi_2$. It is here that symmetry considerations play a role. For $\exp i\mathbf{r}_1\cdot\mathbf{r}_2$ is invariant under a rotation of axes; that is, it commutes with $\mathbf{L}$, given by

$$\mathbf{L} = \boldsymbol{l}_1 + \boldsymbol{l}_2.$$

Since $C_m^{(l)}(\theta_1\phi_1)$ and $C_{m'}^{(l')}(\theta_2\phi_2)$ are tensors of ranks l and l' with respect to $\mathbf{L}$, they can only occur in the coupled scalar form $\mathbf{C}_1^{(l)}\cdot\mathbf{C}_2^{(l)}$. This means that the expansion must run

$$\exp i\mathbf{r}_1\cdot\mathbf{r}_2 = \sum_l a_l j_l(r_1r_2)(\mathbf{C}_1^{(l)}\cdot\mathbf{C}_2^{(l)}).$$

It is highly convenient that both radial dependences can be represented through a common spherical Bessel function.

It remains to find the coefficients a_l. If we take the special case $\theta_1 = \theta_2 = 0$, $\phi_1 = \phi_2$, we get

$$e^{ix} = \sum_l a_l j_l(x). \qquad (5.2)$$

It can be readily verified that the series expansion

$$j_l(x) = \sum_{n=l}^{\infty} \frac{(-1)^{n+l}2^l n!}{(2n+1)!(n-l)!} x^{2n-l} \qquad (5.3)$$

satisfies Eq. (5.1); at the same time, it serves to normalize the functions $j_l(x)$ in the usual way. Since the series for $j_l(x)$ begins with x^l, it is a simple matter to expand both sides of Eq. (5.2) and equate coefficients to find the initial terms. We get

$$a_0 = 1, \quad a_1 = 3i, \quad a_2 = -5, \quad a_3 = -7i, \ \ldots, \qquad (5.4)$$

which suggests the general solution

$$a_l = (2l+1)i^l. \qquad (5.5)$$

To verify that this is in fact the case, the recurrence relation

$$(2l+1)\frac{dj_l(x)}{dx} = lj_{l-1}(x) - (l+1)j_{l+1}(x),$$

which can be derived from Eq. (5.3), is used to obtain

$$\frac{l+1}{2l+3} a_{l+1} - \frac{l}{2l-1} a_{l-1} = ia_l$$

by differentiating both sides of Eq. (5.2). This relates a_{l+1} to the two earlier members of the series, and is consistent with Eqs. (5.4)–(5.5). We can therefore conclude that

$$\exp i\mathbf{r}_1\cdot\mathbf{r}_2 = \sum_l (2l+1)i^l j_l(r_1 r_2)(\mathbf{C}_1^{(l)}\cdot\mathbf{C}_2^{(l)}). \qquad (5.6)$$

In the form for which $r_1 = k$ and $r_2 = r$, this equation corresponds to the expansion of a plane wave in partial waves of well-defined l. An alternative derivation of Eq. (5.6) has been given by Talman [29].

5.2 EXPRESSIONS FOR ELECTRIC POTENTIALS

As a second example, we consider the single-center expansion of V, defined by

$$V = r_{12}^{-l-1} C_m^{(l)}(\theta_{12}\phi_{12}),$$

where $\mathbf{r}_{12} = \mathbf{r}_1 - \mathbf{r}_2$. As in the previous example, the angular variables of $\mathbf{r}_1$ and $\mathbf{r}_2$ are supposed to enter through the harmonic functions $C_q^{(k)}$. The radial functions that attach to a particular $C_q^{(k)}$ are very easy to find, since

$$\nabla_1^2 V = \nabla_2^2 V = 0.$$

In fact, we know from electrostatic theory that two forms are acceptable, namely

$$r^k C_q^{(k)}(\theta\phi), \qquad \text{and} \qquad r^{-k-1} C_q^{(k)}(\theta\phi).$$

For convergence, the expansion must take the form

$$r_{12}^{-l-1} C_m^{(l)}(\theta_{12}\phi_{12}) = \sum_{kk'qq'} r_1^k r_2^{-k'-1} A(kk'qq') C_q^{(k)}(\theta_1\phi_1) C_{q'}^{(k')}(\theta_2\phi_2) \tag{5.7}$$

when $r_1 < r_2$; a similar term is needed to cope with the case $r_1 > r_2$. The coefficients $A(kk'qq')$ are as yet undetermined.

We again introduce the angular momentum vector $\mathbf{L}$, given by $\mathbf{l}_1 + \mathbf{l}_2$. The function r_{12}^{-l-1} is obviously scalar with respect to $\mathbf{L}$. As for $C_m^{(l)}(\theta_{12}\phi_{12})$, we can conveniently regard this as a function of the direction cosines

$$\frac{x_1 - x_2}{r_{12}}, \qquad \frac{y_1 - y_2}{r_{12}}, \qquad \frac{z_1 - z_2}{r_{12}},$$

for these satisfy commutation relations with respect to $\mathbf{L}$ that parallel exactly those satisfied by $(x_1/r_1, y_1/r_1, z_1/r_1)$ with respect to $\mathbf{l}_1$. For example,

$$[L_x, (y_1 - y_2)/r_{12}] = i(z_1 - z_2)/r_{12},$$
$$[l_{1x}, y_1/r_1] = iz_1/r_1.$$

Thus $C_m^{(l)}(\theta_{12}\phi_{12})$ must satisfy the same commutation relations with respect to $\mathbf{L}$ that $C_m^{(l)}(\theta_1\phi_1)$ does with respect to $\mathbf{l}_1$; in other words, it is the component m of a tensor of rank l. Thus the harmonics $\mathbf{C}^{(k)}$ and $\mathbf{C}^{(k')}$ of Eq. (5.7) must be coupled to rank l. Since consistency in the radial dimensions demands $k' = l + k$, the coupling of k and k' to l takes the stretched form

$$(\mathbf{C}_1^{(k)}\mathbf{C}_2^{(l+k)})^{(l)}.$$

We can now write

$$r_{12}^{-l-1} C_m^{(l)}(\theta_{12}\phi_{12}) = \sum_k r_1^k r_2^{-k-l-1} B(k) (\mathbf{C}_1^{(k)}\mathbf{C}_2^{(k+l)})^{(l)}$$
$$+ \sum_k r_2^k r_1^{-k-l-1} B'(k) (\mathbf{C}_1^{(k+l)}\mathbf{C}_2^{(k)})^{(l)},$$

where the first sum is to be taken for $r_1 < r_2$, and the second for $r_1 > r_2$.

The coefficients $B(k)$ and $B'(k)$ can be easily found by taking

$$\theta_1 = \theta_2 = \phi_1 = \phi_2 = \phi_{12} = 0.$$

If $r_1 < r_2$, then $\theta_{12} = \pi$. Setting $m = 0$, we get

$$(-1)^l/(r_2 - r_1)^{l+1} = \sum_k r_1^k r_2^{-k-l-1} B(k)\,(k0,\, k+l\,0 \mid l0),$$

since only the components for which $q = q' = 0$ can contribute. The factor $(-1)^l$ comes from $C_0^{(l)}(\pi, 0)$. The CG coefficient can be explicitly evaluated (see Appendix 1):

$$(k0,\, k+l\,0 \mid l0) = (-1)^l(2l+1)^{1/2}\begin{pmatrix} k & k+l & l \\ 0 & 0 & 0 \end{pmatrix}$$

$$= (-1)^k \frac{(l+k)!}{l!k!}\sqrt{\frac{(2l+1)!(2k)!}{(2l+2k+1)!}}.$$

On the other hand,

$$(r_2 - r_1)^{-l-1} = \sum_k r_1^k r_2^{-l-k-1}(l+k)!/l!k!,$$

so we can deduce that

$$B(k) = (-1)^{k+l}\{(2l+2k+1)!/(2l+1)!(2k)!\}^{1/2}.$$

In the case of $r_1 > r_2$ (for which $\theta_{12} = 0$), we take $r_{12} = r_1 - r_2$, thereby getting $B'(k) = (-1)^l B(k)$. Hence we can conclude that

$$\frac{\mathbf{C}^{(l)}(\theta_{12}\phi_{12})}{r_{12}^{l+1}} = \sum_k (-1)^k \sqrt{\frac{(2l+2k+1)!}{(2l+1)!(2k)!}}\left[\frac{r_2^k}{r_1^{k+l+1}}\left(\mathbf{C}_1^{(k+l)}\mathbf{C}_2^{(k)}\right)^{(l)} + (-1)^l \frac{r_1^k}{r_2^{k+l+1}}\left(\mathbf{C}_1^{(k)}\mathbf{C}_2^{(k+l)}\right)^{(l)}\right]. \tag{5.8}$$

The first term in the brackets is to be used for $r_1 > r_2$, the second for $r_1 < r_2$. Only one term is nonzero at a time. Equation (2.42) is a special case of this result corresponding to $l = 0$. Examples for $l = 1$ and $l = 2$ occur in the analysis of the spin-other-orbit interaction and the spin-spin interaction respectively [16]. Expansions of $r_{12}^n\mathbf{C}^{(l)}(\theta_{12}\phi_{12})$ for cases other than $n = l$ or $n = -l - 1$ can be obtained by recoupling techniques. The results are generally more complex than Eq. (5.8) (see, for example, Problem 5.2).

5.3 DELTA FUNCTIONS

The expansion just obtained for $r_{12}^{-l-1}C_m^{(l)}(\theta_{12}\phi_{12})$ divides space into the two regions $r_1 > r_2$ and $r_1 < r_2$. The case for $r_1 = r_2$ is excluded. One

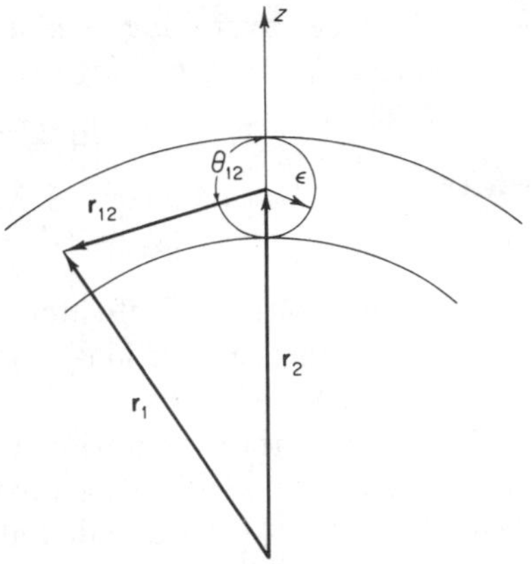

Fig. 5.1. Section containing the vectors $\mathbf{r}_1$, $\mathbf{r}_2$, and $\mathbf{r}_{12}$ for which $\mathbf{r}_1$ lies in the thin shell of thickness 2ϵ.

way to examine the validity of this procedure is to imagine r_2 fixed and to allow r_1 to run up to $r_2 - \epsilon$ and from $r_2 + \epsilon$. The spherical shell of thickness 2ϵ contains the region for which $r_1 = r_2$, which can be studied in the limit $\epsilon \to 0$. This has been done by Pitzer *et al.* [69], who concluded that for sufficiently large values of l the shell produces nonvanishing contributions.

It is convenient to set $\theta_2 = 0$, so that $\mathbf{r}_2$ lies on the z axis. The properties of the function in question inside the thin shell can be studied by choosing functions $f(\mathbf{r}_{12})$ (regular at $\mathbf{r}_{12} = 0$) and calculating

$$I = \int r_{12}^{-l-1} C_m^{(l)}(\theta_{12}\phi_{12}) f(\mathbf{r}_{12})\, d\tau_{12}, \tag{5.9}$$

the integration extending over the volume of the shell. For example, if I is found to contain a term in $f(0)$, then a simple delta function is present; if $f'(0)$ occurs, then derivatives of the delta function are needed. As indicated in Fig. 5.1, the thin shell is broken up into the two regions $r_{12} \geq \epsilon$ and $\epsilon > r_{12} \geq 0$. The second region is a small sphere of radius ϵ, for which θ_{12} and ϕ_{12} can run over their entire ranges $0 \leq \phi_{12} \leq 2\pi$ and $0 \leq \theta_{12} \leq \pi$. The harmonic $C_m^{(l)}$ in I integrates to zero within this sphere unless $f(\mathbf{r}_{12})$ contains, in its expansion near $\mathbf{r}_{12} = 0$, a harmonic of identical rank. For this to be possible, l differentiations of $f(\mathbf{r}_{12})$ are needed, and the associated power of r_{12} in the Taylor expansion is also l. However,

$$\lim_{\epsilon \to 0} \int_0^{\epsilon} r_{12}^{-l-1} \cdot r_{12}^{l} \cdot r_{12}^{2}\, dr_{12} = 0,$$

so even in this case the contribution from the small sphere is nil.

We are left with the region $r_{12} \geq \epsilon$. The integration over ϕ_{12} is straightforward, since, by taking $\mathbf{r}_2$ along the z axis, the range $0 \leq \phi_{12} \leq 2\pi$ is retained. From the geometry of Fig. 5.1, we see that

$$\mu_+ \geq \cos\theta_{12} \geq \mu_-,$$

where

$$\mu_\pm = (\epsilon^2 \pm 2r_2\epsilon - r_{12}^2)/2r_2r_{12}.$$

For r_{12}, an arbitrary finite upper bound $r_{12} = r_0$ can be applied to the integration; all contributions from this boundary go to zero as $\epsilon \to 0$, as will be immediately seen.

To visualize the integration and subsequent limiting process with as much clarity as possible, it seems best to take an example. We select $l = 3$. The function $f(\mathbf{r}_{12})$ is expanded about $\mathbf{r}_{12} = 0$:

$$f(\mathbf{r}_{12}) = f(0) + [(\mathbf{r}_{12}\cdot\nabla_{12})f(\mathbf{r}_{12})]_{\mathbf{r}_{12}=0} + \cdots. \tag{5.10}$$

Retaining only those terms that can contribute, we find

$$\int r_{12}^{-4}C_0^{(3)}(\theta_{12}\phi_{12})f(0)\,d\tau_{12} = -\frac{\pi}{r_2}f(0)\left[5\frac{\epsilon^5}{r_{12}^6} - 8\frac{\epsilon^3}{r_{12}^4} + 3\frac{\epsilon}{r_{12}^2}\right]_\epsilon^{r_0}$$

$$= (4\pi/3r_2)f(0)$$

in the limit $\epsilon \to 0$. Thus I contains the delta function $4\pi\delta(\mathbf{r}_1 - \mathbf{r}_2)/3r_2$. The contributions coming from the bracketed term of Eq. (5.10) can be calculated in a similar way. For example,

$$\int r_{12}^{-4}C_0^{(3)}(\theta_{12}\phi_{12})r_{12}\cos\theta_{12}\left[\frac{\partial f(\mathbf{r}_{12})}{\partial z_{12}}\right]_{r_{12}=0} d\tau_{12} = -\frac{4\pi}{15}\left[\frac{\partial f(\mathbf{r}_{12})}{\partial z_{12}}\right]_{r_{12}=0}.$$

This result can be reproduced by $4\pi\delta_0^{(1)}(\mathbf{r}_1 - \mathbf{r}_2)/15$, where the vector operator $\boldsymbol{\delta}^{(1)}$ is defined by

$$\boldsymbol{\delta}^{(1)}(\mathbf{r}) = \nabla\delta(\mathbf{r}).$$

The reversal of sign is necessary because

$$\int f(\mathbf{r})\delta_0^{(1)}(\mathbf{r})\,d\mathbf{r} = \int f(\mathbf{r})\frac{\partial}{\partial z}\delta(\mathbf{r})\,dx\,dy\,dz$$

$$= -\int\frac{\partial f(\mathbf{r})}{\partial z}\delta(\mathbf{r})\,dx\,dy\,dz$$

by partial integration. All in all, the delta-function contributions to $r_{12}^{-4}C_m^{(3)}(\theta_{12}\phi_{12})$ are found to be

$$(4\pi/3r_2)\delta(\mathbf{r}_1 - \mathbf{r}_2) + (4\pi/15)\delta_0^{(1)}(\mathbf{r}_1 - \mathbf{r}_2)$$

for $m = 0$, and

$$(4\pi/15)(6)^{1/2}\delta_{\pm 1}^{(1)}(\mathbf{r}_1 - \mathbf{r}_2)$$

for $m = \pm 1$.

To express these results in a general way, the condition $\theta_2 = 0$ must be relaxed. Since the overall rank is 3, the general form for the terms must be

$$a\mathbf{C}^{(3)}(\theta_2\phi_2)\delta(\mathbf{r}_1 - \mathbf{r}_2) + b[\mathbf{C}^{(2)}(\theta_2\phi_2)\boldsymbol{\delta}^{(1)}(\mathbf{r}_1 - \mathbf{r}_2)]^{(3)} + c[\mathbf{C}^{(4)}(\theta_2\phi_2)\boldsymbol{\delta}^{(1)}(\mathbf{r}_1 - \mathbf{r}_2)]^{(3)}. \quad (5.11)$$

No term of the type $(\mathbf{C}^{(3)}\boldsymbol{\delta}^{(1)})^{(3)}$ is required, since this possesses the wrong parity with respect to the simultaneous operations $\mathbf{r}_1 \to -\mathbf{r}_1$ and $\mathbf{r}_2 \to -\mathbf{r}_2$. To find the coefficients a, b, and c, we simply insist that the special cases for $\theta_2 = 0$ be reproduced. In this way we get

$$a = (4\pi/3r_2), \qquad b = (4\pi/21)(15)^{1/2}, \qquad c = 16\pi/35. \quad (5.12)$$

These numbers serve to define the entire delta-function component of $r_{12}^{-4}\mathbf{C}^{(3)}(\theta_{12}\phi_{12})$.

No delta functions are required for $l = 0$ or 1. For $l = 2$, the single term

$$-(4\pi/3)\mathbf{C}^{(2)}(\theta_2\phi_2)\delta(\mathbf{r}_1 - \mathbf{r}_2)$$

is required. Kay *et al.* [70] have examined the general case by different methods from those described here (see Problem 5.4).

5.4 BIPOLAR EXPANSIONS

The presence in molecules of more than one nucleus makes it important for us to be able to carry out expansions about two centers (A and B say) simultaneously. It is convenient to label two arbitrary points by the numerals 1 and 2; their positions with respect to A and B are specified by such vectors as $\mathbf{r}_{1A}$. The pattern of vectors is illustrated in Fig. 5.2. In general, we define $\mathbf{r}_{XY} = \mathbf{r}_X - \mathbf{r}_Y$, where X and Y can be letters or numerals. In addition, the direction of $\mathbf{r}_{XY}$ is supposed to be defined by the polar angles $(\theta_{XY}\phi_{XY})$ in the usual way:

$$x_X - x_Y = r_{XY}\sin\theta_{XY}\cos\phi_{XY},$$

$$y_X - y_Y = r_{XY}\sin\theta_{XY}\sin\phi_{XY},$$

$$z_X - z_Y = r_{XY}\cos\theta_{XY}, \quad (5.13)$$

where $r_{XY} = |\mathbf{r}_{XY}|$.

The separation of the radial space, which, for the single-center expansion of Section 5.2, corresponds to taking $r_1 > r_2$, $r_1 = r_2$, or $r_1 < r_2$, now

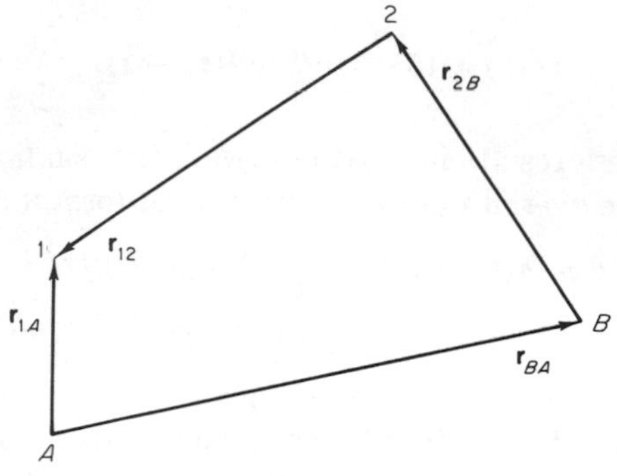

Fig. 5.2. Vectors for expansions about the two centers A and B.

becomes more complicated. Buehler and Hirschfelder [71] have introduced a very useful diagram for this purpose: it is sketched in Fig. 5.3. To illustrate our techniques, we shall limit attention to the expansion of r_{12}^{-1} for region I, for which $r_{1A} + r_{2B} < r_{AB}$. This corresponds to the electrostatic interaction between the electrons 1 and 2 on two atoms whose nuclear separation r_{AB} is so great that overlap is negligible.

The general method can be applied as before. If we wish to obtain an expansion in which $\mathbf{r}_{1A}$ and $\mathbf{r}_{2B}$ enter, then the relation

$$\mathbf{r}_{12} = \mathbf{r}_{1A} - \mathbf{r}_{2B} + \mathbf{r}_{AB} \tag{5.14}$$

shows that we must introduce quantities depending on $\mathbf{r}_{AB}$. It also shows that if two of the three radii on the right-hand side are kept fixed—say $\mathbf{r}_{1A}$ and $\mathbf{r}_{AB}$—then the Laplacian ∇_{12}^2 constructed from the coordinates $\mathbf{r}_{12}$ is identical to ∇_{2B}^2. Since $\nabla_{12}^2(r_{12}^{-1}) = 0$ (for region I, where $r_{12} > 0$), the functions of $\mathbf{r}_{1A}$, $\mathbf{r}_{2B}$, and $\mathbf{r}_{AB}$ that we use in the expansion of r_{12}^{-1} must all separately satisfy Laplace's equation. That is, the function $\psi(\mathbf{r}_{XY})$ of $\mathbf{r}_{XY}$ must satisfy

$$\nabla_{XY}^2\psi(\mathbf{r}_{XY}) = 0.$$

It follows that the dependence of r_{12}^{-1} on $\mathbf{r}_{1A}$ is represented by some combination of harmonics of the type $r_{1A}{}^k\mathbf{C}_{1A}{}^{(k)}$ or $r_{1A}{}^{-k-1}\mathbf{C}_{1A}{}^{(k)}$, where, in general,

$$\mathbf{C}_{XY}{}^{(k)} \equiv \mathbf{C}^{(k)}(\theta_{XY}\phi_{XY}).$$

Now, for region I,

$$r_{1A} < r_{AB} - r_{2B} < r_{2A}.$$

Thus electron 1 is nearer to A than electron 2, and hence harmonics of the first kind only are required. Similarly, $r_{2B} < r_{1B}$, and so we only need

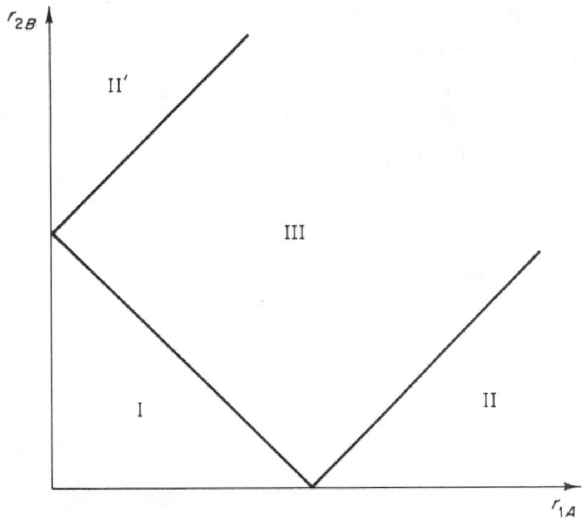

Fig. 5.3. The Buehler–Hirschfelder diagram. The radial space for a bipolar expansion is divided into the four regions I, II, II', and III for which the following inequalities hold:

$$I:\quad r_{1A} + r_{2B} < r_{AB},$$

$$II:\quad r_{2B} + r_{AB} < r_{1A},$$

$$II':\quad r_{1A} + r_{AB} < r_{2B},$$

$$III:\quad |r_{1A} - r_{2B}| < r_{AB} < r_{1A} + r_{2B}.$$

In addition to these four regions, the lines of demarcation often require separate analysis, since they may give rise to delta functions in the expansion.

harmonics of the type $r_{2B}{}^k C_{2B}{}^{(k)}$ to express the dependence of r_{12}^{-1} on $\mathbf{r}_{2B}$. Its dependence on $\mathbf{r}_{AB}$, on the other hand, is represented by the harmonics $r_{AB}{}^{-k-1}\mathbf{C}_{AB}{}^{(k)}$, since r_{AB} has a lower bound, but not an upper one.

We can now assemble the various parts. However, it is a simple matter to give the resulting expansion a slightly more precise structure by using the fact that every term in it must, like r_{12}^{-1}, possess the dimensions of an inverse length. Thus we get

$$r_{12}^{-1} = \sum_{lkmqp} A(lkmqp) r_{1A}{}^l (C_m{}^{(l)})_{1A} r_{2B}{}^k (C_q{}^{(k)})_{2B} r_{AB}{}^{-l-k-1} (C_p{}^{(l+k)})_{AB}. \quad (5.15)$$

To progress further, we introduce the following angular momentum vector:

$$\mathbf{L} = \boldsymbol{l}_{1A} + \boldsymbol{l}_{2B} + \boldsymbol{l}_{AB},$$

where $\boldsymbol{l}_{XY}$ is the same function of θ_{XY} and ϕ_{XY} as $\boldsymbol{l}$ (of Eqs. (1.6)) is of θ and ϕ. We note in passing that the truncated form $\boldsymbol{l}_{1A} + \boldsymbol{l}_{2B}$ has been used by Copland and Newman [72] in their studies of superexchange.

The three sets of angle pairs $(\theta_{1A}\phi_{1A})$, $(\theta_{2B}\phi_{2B})$, and $(\theta_{AB}\phi_{AB})$ can be regarded as independent variables, so

$$[\mathbf{l}_{1A}, \mathbf{l}_{2B}] = [\mathbf{l}_{2B}, \mathbf{l}_{AB}] = [\mathbf{l}_{AB}, \mathbf{l}_{1A}] = 0.$$

To test r_{12}^{-1} under the action of $\mathbf{L}$, we write

$$r_{12}^{-1} = \{(x_{1A} - x_{2B} + x_{AB})^2 + (y_{1A} - y_{2B} + y_{AB})^2 + (z_{1A} - z_{2B} + z_{AB})^2\}^{-1/2}.$$

Then it is easy to show that

$$\begin{aligned}[L_x, r_{12}^{-1}] &= [(l_x)_{1A}, r_{12}^{-1}] + [(l_x)_{2B}, r_{12}^{-1}] + [(l_x)_{AB}, r_{12}^{-1}] \\ &= i r_{12}^{-3}\{y_{1A}(z_{1A} - z_{2B} + z_{AB}) - z_{1A}(y_{1A} - y_{2B} + y_{AB}) \\ &\quad + y_{2B}(-z_{1A} + z_{2B} - z_{AB}) - z_{2B}(-y_{1A} + y_{2B} - y_{AB}) \\ &\quad + y_{AB}(z_{1A} - z_{2B} + z_{AB}) - z_{AB}(y_{1A} - y_{2B} + y_{AB})\} \\ &= 0.\end{aligned}$$

By cyclic permutation, r_{12}^{-1} commutes with every component of $\mathbf{L}$. The three harmonics of Eq. (5.15) must thus be coupled to a rank of zero. There is only one way of doing this:

$$r_{12}^{-1} = \sum_{lk} B(lk) r_{1A}{}^{l} r_{2B}{}^{k} r_{AB}{}^{-l-k-1} (\mathbf{C}_{1A}{}^{(l)}\mathbf{C}_{2B}{}^{(k)})^{(l+k)} \cdot \mathbf{C}_{AB}{}^{(l+k)}. \qquad (5.16)$$

To fix the coefficients $B(lk)$, we pick the special case $\theta_{1A} = \theta_{2B} = \theta_{AB} = \pi$, for which $\mathbf{r}_{1A}$, $\mathbf{r}_{2B}$, and $\mathbf{r}_{AB}$ become aligned, with

$$\begin{aligned}r_{12}^{-1} &= (r_{AB} + r_{1A} - r_{2B})^{-1} \\ &= \sum_{lk} r_{1A}{}^{l} r_{2B}{}^{k} r_{AB}{}^{-l-k-1} (-1)^l (l + k)!/l!k!\end{aligned}$$

The scalar product on the extreme right of Eq. (5.16) reduces to a CG coefficient of the same type as the one that arose in Section 5.2. It is now easy to evaluate $B(lk)$, and Eq. (5.16) becomes

$$\frac{1}{r_{12}} = \sum_{lk} \frac{r_{1A}{}^{l} r_{2B}{}^{k}}{r_{AB}{}^{l+k+1}} \sqrt{\frac{(2l + 2k)!}{(2l)!(2k)!}} [(\mathbf{C}_{A1}{}^{(l)}\mathbf{C}_{2B}{}^{(k)})^{(l+k)} \cdot \mathbf{C}_{AB}{}^{(l+k)}]. \qquad (5.17)$$

The parity relation $C_{A1}{}^{(l)} = (-1)^l C_{1A}{}^{(l)}$ has been used here to remove a phase factor. This expansion for r_{12}^{-1} was first given (in an uncoupled form, however) by Carlson and Rushbrooke [73]. More complicated bipolar expansions have been studied by Kay *et al.* [70, p. 2363] using the Fourier transform techniques of Section 5.6. (See also Problem 5.5.) It is important to appreciate that formulae such as Eqs. (5.8) and (5.17) should not be put in an uncoupled form before it is absolutely necessary. The angular momentum techniques of Chapter 1 are designed to avoid uncouplings as far as is possible. This is to achieve maximum mathematical compression.

5.5 THE NEUMANN EXPANSION

In spite of their somewhat awkward nature, spheroidal coordinates can be used to give a remarkably simple expansion for r_{12}^{-1}. The original derivation, due to Neumann [74], has been recast in a more straightforward form by Ruedenberg [75]. The derivation to be given here relies more on symmetry to account for the simple structure of the result.

We start by working out the analogs of the harmonics $r^k\mathbf{C}^{(k)}$ and $r^{-k-1}\mathbf{C}^{(k)}$. The Laplacian ∇^2 has been given in spheroidal coordinates in Eq. (4.8). Written slightly differently, it runs

$$\nabla^2 = b^{-2}(\xi^2 - \eta^2)^{-1}\{\Delta(\eta, \phi) - \Delta(\xi, \phi)\}, \tag{5.18}$$

where

$$\Delta(\mu, \phi) \equiv \frac{\partial}{\partial\mu}\left\{(1 - \mu^2)\frac{\partial}{\partial\mu}\right\} + \frac{1}{1 - \mu^2}\frac{\partial^2}{\partial\phi^2}.$$

This operator appears in the familiar equation satisfied by spherical harmonics, namely [76]

$$\{\Delta(\mu, \phi) + l(l + 1)\}P_l^m(\mu)e^{im\phi} = 0. \tag{5.19}$$

It is now clear that we can construct a solution to Laplace's equation in spheroidal coordinates by taking

$$U_l^m(\xi\eta\phi) = P_l^m(\xi)P_l^m(\eta)e^{im\phi}, \tag{5.20}$$

since

$$\begin{aligned}
&\{\Delta(\eta, \phi) - \Delta(\xi, \phi)\}P_l^m(\xi)P_{l'}^m(\eta)e^{im\phi} \\
&\quad = P_l^m(\xi)\{\Delta(\eta, \phi)P_{l'}^m(\eta)e^{im\phi}\} - P_{l'}^m(\eta)\{\Delta(\xi, \phi)P_l^m(\xi)e^{im\phi}\} \\
&\quad = \{l(l + 1) - l'(l' + 1)\}P_l^m(\xi)P_{l'}^m(\eta)e^{im\phi},
\end{aligned} \tag{5.21}$$

and this vanishes when $l' = l$.

The harmonics U_l^m do not exhaust all solutions to Laplace's equation. As ξ tends to infinity, we find

$$P_l^m(\xi) \to \xi^l(2l)!/2^l l!(l - m)! \tag{5.22}$$

In the limit of large ξ, we can make the replacement $\xi \to r/b$, so the harmonics U_l^m are the analogs of $r^lC_m^{(l)}$. To find the harmonics that correspond to $r^{-l-1}C_m^{(l)}$, we use the fact that Eq. (5.19) remains valid if P_l^m is replaced by Q_l^m, where Q is an associated Legendre function of the second kind. Its form can most easily be found by making the replacement $l \rightarrow -l - 1$ in Eq. (4.25) and interpreting an inverse differentiation as an integration. Putting in the conventional prefacing coefficient, we get

$$Q_l^m(\xi) = 2^l l!(\xi^2 - 1)^{m/2} \int_\xi^\infty \cdots \int_\xi^\infty (\xi^2 - 1)^{-l-1} (d\xi)^{l+1-m} \qquad (\xi > 1). \tag{5.23}$$

The substitution $l \rightarrow -l - 1$ leaves the characteristic form $l(l + 1)$ invariant, and has been extensively studied by Jucys and his collaborators in a different context [2]. In the limit $\xi \rightarrow \infty$, we find

$$Q_l^m(\xi) \rightarrow \xi^{-l-1} 2^l l!(l + m)!/(2l + 1)!,$$

so the expected proportionality to r^{-l-1} appears. We write

$$V_l^m(\xi\eta\phi) = Q_l^m(\xi) P_l^m(\eta) e^{im\phi} \tag{5.24}$$

for the second class of spheroidal harmonics.

Both kinds of spheroidal harmonics are available for expressing the electric potential produced by a charge distribution: their roles parallel those of the two polar harmonics. For example, if a charge distribution is located outside the ellipsoid $\xi = a$, then the potential inside can be expressed as a superposition of the harmonics U_l^m. Conversely, the harmonics V_l^m are appropriate for representing the external potential produced by a system of charges lying wholly within the ellipsoid $\xi = a$. This is a matter of electrostatics [56].

We can regard r_{12}^{-1} from two points of view. On the one hand, it represents the potential at $\mathbf{r}_2$ produced by a charge situated at $\mathbf{r}_1$. If, for definiteness, we suppose $\xi_1 > \xi_2$, then we can write

$$r_{12}^{-1} = \sum_{kq} A(kq) U_k^q(\xi_2\eta_2\phi_2),$$

where the coefficients $A(kq)$ depend on $\mathbf{r}_1$. On the other hand, r_{12}^{-1} can be thought of as the potential at $\mathbf{r}_1$ produced by a charge at $\mathbf{r}_2$. In this case,

$$r_{12}^{-1} = \sum_{lm} B(lm) V_l^m(\xi_1\eta_1\phi_1),$$

where the coefficients $B(lm)$ depend on $\mathbf{r}_2$. These two expansions are now put together. Since ϕ_1 and ϕ_2 can only appear through the combination $\phi_1 - \phi_2$, the sum over q can be removed, and we get, in full detail,

$$r_{12}^{-1} = \sum_{klm} C(klm) Q_l^m(\xi_1) P_l^m(\eta_1) P_k^{-m}(\xi_2) P_k^{-m}(\eta_2) \exp im(\phi_1 - \phi_2), \tag{5.25}$$

where the coefficients $C(klm)$ are independent of $\mathbf{r}_1$ and $\mathbf{r}_2$. An expansion of this form is valid for any axially symmetric function ψ for which both $\nabla_1^2\psi$ and $\nabla_2^2\psi$ are zero. What distinguishes r_{12}^{-1} from other solutions—and, incidentally, makes it possible to continue the analysis—is the fact that it is invariant under the interchange $\eta_1 \leftrightarrow \eta_2$. It is probably easiest to see this by writing out r_{12}^2 explicitly:

$$(r_{12}/b)^2 = \xi_1^2 + \eta_1^2 + \xi_2^2 + \eta_2^2 - 2\xi_1\eta_1\xi_2\eta_2 - 2$$
$$- 2\{(\xi_1^2 - 1)(1 - \eta_1^2)(\xi_2^2 - 1)(1 - \eta_2^2)\}^{1/2}\cos(\phi_1 - \phi_2).$$

Thus

$$r_{12}^{-1} = \sum_{klm} C(klm) Q_l^m(\xi_1) P_k^{-m}(\eta_1) P_k^{-m}(\xi_2) P_l^m(\eta_2) \exp im(\phi_1 - \phi_2).$$

If, now, we evaluate $\nabla_2^2(r_{12}^{-1})$, we find

$$0 = \sum_{klm} C(klm) Q_l^m(\xi_1) P_k^{-m}(\eta_1) P_k^{-m}(\xi_2) P_l^m(\eta_2) \exp im(\phi_1 - \phi_2)$$
$$\times \{k(k+1) - l(l+1)\}. \tag{5.26}$$

To get this result, we use Eqs. (5.18)–(5.19) in precisely the same way as they were used to obtain Eq. (5.21). We can now take advantage of the appearance of the independent coordinates η_1, η_2, and ϕ_2 in Eq. (5.26) to equate each term of the sum in this equation to zero. The conclusion is that either $C(klm) = 0$ or $k = l$. The triple sum in Eq. (5.26) thus reduces to a sum over l and m.

It only remains to find $C(llm)$ by ensuring that the expansion goes over into the familiar form of Eq. (2.42) as ξ_1 and ξ_2 tend to infinity. In this way, we get

$$C(llm) = (-1)^m(2l + 1)/b.$$

Before writing down the final result, we remove the condition $\xi_1 > \xi_2$ by simply making the replacements $\xi_1 \to \xi_>$ and $\xi_2 \to \xi_<$, where $\xi_>$ and $\xi_<$ are the greater and lesser respectively of ξ_1 and ξ_2. Because r_{12} is invariant under the interchange $\xi_1 \leftrightarrow \xi_2$, there is no need to adjust the subscripts to η and ϕ. The conclusion is

$$r_{12}^{-1} = b^{-1} \sum_{lm} (2l + 1)(-1)^m Q_l^m(\xi_>) P_l^m(\eta_1) P_l^{-m}(\xi_<) P_l^{-m}(\eta_2)$$
$$\times \exp im(\phi_1 - \phi_2), \tag{5.27}$$

the so-called Neumann expansion. Slater [77] has shown how it can be applied to evaluate the Coulomb matrix elements involving 1s hydrogenic eigenfunctions centered on two separated nuclei. His discussion gives more detail than the original article of Sugiura [78] on the hydrogen molecule.

If the U and V harmonics are used, Eq. (5.27) becomes

$$r_{12}^{-1} = b^{-1} \sum_{lm} (2l+1)(-1)^m U_l^{-m}(\xi_< \eta_2 \phi_2) V_l^m(\xi_> \eta_1 \phi_1). \tag{5.28}$$

The sum over m strongly suggests that some kind of scalar product is being formed between quantities that appear to be the components of tensors. However, it is not difficult to show that, from the simple differential operators $\partial/\partial\xi$, $\partial/\partial\eta$, and $\partial/\partial\phi$, we can construct no linear combinations that act as the generators of an $R(3)$ group for which the U and V functions play the role of tensors. The characteristically scalar appearance of Eq. (5.28) seems to come not from an $R(3)$ symmetry but rather from the invariance of r_{12}^{-1} under the interchange $\eta_1 \leftrightarrow \eta_2$. Support for this view comes from the work of Matcha *et al.* [79], who have obtained the expansions in spheroidal harmonics of more complicated solutions to Laplace's equation such as $r_{12}^{-3}\mathbf{C}^{(2)}(\theta_{12}\phi_{12})$. Their expansions do not reveal any obvious coupling between the U and V functions; nor would any have been expected were it not for the peculiar simplicity of Eq. (5.28).

5.6 RUEDENBERG'S METHOD

An ingenious technique for deriving the kind of expansions discussed in previous sections has been described by Ruedenberg [80]. To show how it works, we derive again the single-center expansion of r_{12}^{-1} that has already been given in Section 2.10. The starting point is virtually a tautology. We use Eqs. (3.15) and (3.18) to write

$$r_{12}^{-1} = (2\pi)^{-3/2} \int \exp(i(\mathbf{r}_1 - \mathbf{r}_2)\cdot\mathbf{p})\, \psi(\mathbf{p})\, d\mathbf{p}, \tag{5.29}$$

where $\psi(\mathbf{p})$ is the Fourier transform of r^{-1}. That is,

$$\begin{aligned} \psi(\mathbf{p}) &= (2\pi)^{-3/2} \int \exp(-i\mathbf{r}\cdot\mathbf{p}) r^{-1}\, d\mathbf{r} \\ &= (2\pi)^{-3/2} 4\pi/p^2, \end{aligned}$$

from the analysis of Section 3.5. The exponential term in Eq. (5.29) is expanded by using Eq. (5.6) twice:

$$\begin{aligned} \exp i(\mathbf{r}_1 - \mathbf{r}_2)\cdot\mathbf{p} = \sum_{kl} &(2l+1)(2k+1) i^l (-i)^k j_l(r_1 p) j_k(r_2 p) \\ &\times (\mathbf{C}_1^{(l)}\cdot\mathbf{C}_p^{(l)})(\mathbf{C}_2^{(k)}\cdot\mathbf{C}_p^{(k)}), \end{aligned}$$

where $\mathbf{C}_p^{(t)}$ is the tensor whose angular variables depend on the direction of $\mathbf{p}$. A recoupling is now performed:

$$\begin{aligned}(\mathbf{C}_1^{(l)}\cdot\mathbf{C}_p^{(l)})(\mathbf{C}_2^{(k)}\cdot\mathbf{C}_p^{(k)}) &= (-1)^{l+k}[l,k]^{1/2}\{(\mathbf{C}_1^{(l)}\mathbf{C}_p^{(l)})^{(0)}(\mathbf{C}_2^{(k)}\mathbf{C}_p^{(k)})^{(0)}\}^{(0)} \\ &= (-1)^{l+k}[l,k]^{1/2}\sum_t((ll)0,(kk)0,0\mid(lk)t,(lk)t,0) \\ &\qquad\times\{(\mathbf{C}_1^{(l)}\mathbf{C}_2^{(k)})^{(t)}(\mathbf{C}_p^{(l)}\mathbf{C}_p^{(k)})^{(t)}\}^{(0)}.\end{aligned}$$

The recoupling coefficient is evaluated by first converting it to a 9-*j* symbol. The result is

$$((ll)0,(kk)0,0\mid(lk)t,(lk)t,0) = [l,k]^{-1/2}[t]^{1/2}.$$

The tensor $(\mathbf{C}_p^{(l)}\mathbf{C}_p^{(k)})^{(t)}$ can be related to $\mathbf{C}_p^{(t)}$ by means of Eq. (2.30). However, of all the tensors of this type, only the one for which $t = 0$ survives the angular integration of Eq. (5.29). We obtain

$$(\mathbf{C}_1^{(l)}\cdot\mathbf{C}_p^{(l)})(\mathbf{C}_2^{(k)}\cdot\mathbf{C}_p^{(k)}) = \delta(l,k)(2l+1)^{-1}(\mathbf{C}_1^{(l)}\cdot\mathbf{C}_2^{(l)}) + A,$$

where A represents the ineffective terms for which $t \neq 0$. The angular integration yields 4π, and we get

$$r_{12}^{-1} = (2\pi)^{-3}\sum_l(2l+1)(\mathbf{C}_1^{(l)}\cdot\mathbf{C}_2^{(l)})\int_0^\infty j_l(r_1p)j_l(r_2p)(4\pi/p^2)4\pi p^2\,dp. \tag{5.30}$$

Now

$$\begin{aligned}\int j_l(r_1p)j_l(r_2p)\,dp &= \tfrac{1}{2}\pi(r_1r_2)^{-1/2}\int p^{-1}J_{l+1/2}(r_1p)J_{l+1/2}(r_2p)\,dp \\ &= \tfrac{1}{2}\pi(2l+1)^{-1}r_1^lr_2^{-l-1} \qquad (r_1 \le r_2)\end{aligned}$$

or

$$= \tfrac{1}{2}\pi(2l+1)^{-1}r_2^lr_1^{-l-1} \qquad (r_1 \ge r_2).$$

This last step, the evaluation of the integral over the Bessel functions, is performed by means of Eq. (1) in Section 13.42 of the book by Watson [81]. Equation (5.30) now assumes the familiar form

$$r_{12}^{-1} = \sum_l(r_<^l/r_>^{l+1})(\mathbf{C}_1^{(l)}\cdot\mathbf{C}_2^{(l)}).$$

The advantage of this method is that it enables a direct analytical approach to the problem in hand. Possible sources of delta functions (such as might occur when $r_1 = r_2$) cannot be easily overlooked. On the other hand, a successful outcome depends very much on one's skill in handling integrals involving Bessel functions. For complicated cases, it may easily happen that the final product of hypergeometric functions may go unrecognized for the coupled product of 3-*j* or 6-*j* symbols that it is.

It will also have been noticed that the evaluation of the Fourier transform of r^{-1} that is described in Section 3.5 depends on a somewhat arbitrary limiting procedure. Fortunately, the theory of generalized functions [82] permits a consistent and systematic approach to be made to problems of this kind. Elaborate use of these techniques has been made by Kay *et al.* [70, p. 2363] in their treatment of the bipolar expansions of $r_{12}{}^n C_{12}{}^{(l)}$.

5.7 ELECTRIC QUADRUPOLE–QUADRUPOLE INTERACTION

Several applications of the preceding expansions have already been mentioned. However, it is probably worthwhile to give a rather detailed example. Most of the single-center expansions play their most obvious role in atomic rather than molecular theory, so it seems appropriate to pass these over and consider instead a bipolar expansion. Much interest attaches to the use of Eq. (5.17) to calculate the electric quadrupole–quadrupole interaction between rare-earth ions embedded in various crystal lattices. A suitable example for our purposes is that of the crystal $LaCl_3$ in which a certain fraction of the lanthanum ions La^{3+} are replaced by Ce^{3+}. There is a finite probability that two neighboring La^{3+} ions are both replaced by Ce^{3+}, and this makes it possible to study the Ce^{3+}–Ce^{3+} interaction. The point symmetry at a lanthanum site in $LaCl_3$ is C_{3h}: that is, there exists a threefold axis of symmetry and a (horizontal) reflection plane containing the site and perpendicular to this axis. The shortest distance between a pair of lanthanum sites occurs when both lie on the threefold axis. The situation we shall study corresponds to the occupation of such a pair of sites by cerium ions.

The ion Ce^{3+} is particularly simple because it contains a single 4f electron surrounded by a spherically-symmetric closed-shell structure. The ground level is $^2F_{5/2}$, but this is split into three components when the cerium ion is embedded in the crystal lattice. The reasonably high symmetry at the site separates the level $^2F_{5/2}$ into the three doublets characterized by $M = \pm 1/2, \pm 3/2, \pm 5/2$. The doublet $\pm 5/2$ is lowest; the doublets $\pm 1/2$ and $\pm 3/2$ lie 37.5 cm^{-1} and 110 cm^{-1} above it [83]. In the absence of any interaction between the two cerium ions, the combined energy-level pattern is merely the superposition of the two individual schemes. This is illustrated in Fig. 5.4. The total number of independent states is 6×6,

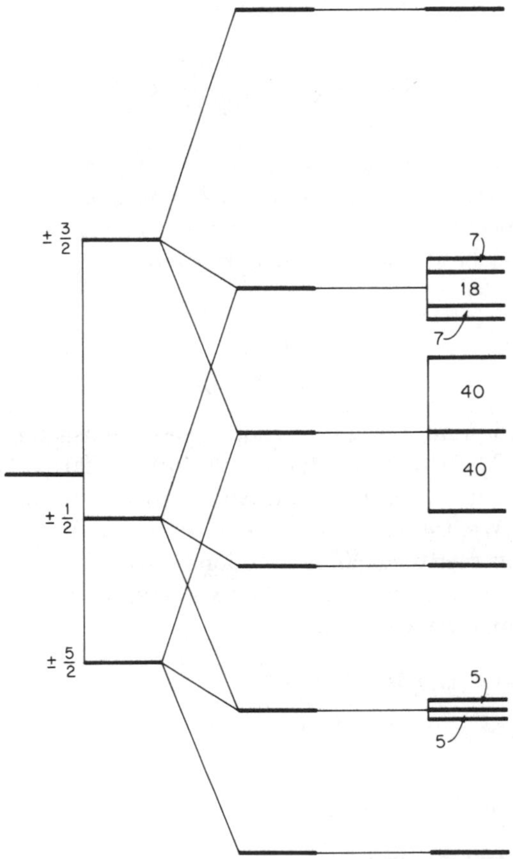

Fig. 5.4. Energy levels for the first-order quadrupole–quadrupole interaction between a pair of Ce^{3+} ions in $LaCl_3$. On the left is the three-doublet structure of the ground level $^2F_{5/2}$ of a single cerium ion. The oblique lines lead to the energy-level scheme of two noninteracting Ce^{3+} ions. The interaction between the quadrupole moment of the 4f electron's charge distribution on one Ce^{3+} ion and that on the other produces the (magnified) splittings on the right. The numbers give the first-order splittings in units of $12e^2\langle r^2\rangle^2/1225r_{AB}^5$, for the situation when the two cerium ions lie on a common threefold crystal axis.

that is, 36; and this multiplicity is distributed among three quartets and three octets. The states are conveniently denoted by $|\,M_1M_2\rangle$.

An obvious choice for the z axis is the threefold axis on which the two cerium ions lie. Their nuclei are assigned the labels A and B of Section 5.4 such that $\theta_{AB} = 0$. For this angle,

$$C_p^{(l+k)}(\theta_{AB}\phi_{AB}) = \delta(p, 0),$$

and Eq. (5.17) runs

$$\frac{1}{r_{12}} = \sum_{lk} \frac{r_{1A}{}^l r_{2B}{}^k}{r_{AB}{}^{l+k+1}} \sqrt{\frac{(2l+2k)!}{(2l)!(2k)!}} (\mathbf{C}_{A1}{}^{(l)}\mathbf{C}_{2B}{}^{(k)})_0{}^{(l+k)}.$$

For the physical system that we are examining, r_{1A} and r_{2B} are much less than r_{AB}. The leading term, for which $l = k = 0$, produces a shift in the total energy spectrum, and is of no interest to us. Terms for which l (or k) is odd can also be dropped because the reduced matrix elements of $\mathbf{C}^{(l)}$ depend on the 3-j symbols

$$\begin{pmatrix} 3 & l & 3 \\ 0 & 0 & 0 \end{pmatrix},$$

as can be seen from Eq. (1.57), and these vanish when l is odd. (See Appendix 1.) The two terms for which $(lk) \equiv (20)$ or (02) merely contribute to the single-ion spectrum which, for us, is the starting point of the analysis. We therefore direct our attention to the term for which $l = k = 2$. This represents an interaction between the quadrupole moments of the f-electron distributions on the two cerium ions.

The basic quantity to calculate is

$$\begin{aligned} &\langle M_1 M_2 \mid e^2/r_{12} \mid M_1' M_2' \rangle \\ &\quad = e^2 r_{AB}{}^{-5} (70)^{1/2} \sum_q (2q, 2\, -q \mid 40) \langle M_1 \mid r_{1A}{}^2 C_q{}^{(2)} \mid M_1' \rangle \\ &\quad\quad \times \langle M_2 \mid r_{2B}{}^2 C_{-q}{}^{(2)} \mid M_2' \rangle. \end{aligned} \tag{5.31}$$

Both matrix elements in the sum can be handled in a similar way. Radial integration for the f electrons can be formally performed by withdrawing $r_{1A}{}^2$ and $r_{2B}{}^2$ from the matrix elements and replacing them by $\langle r^2 \rangle$. Putting in a more complete description for the bras and kets, we get, from the WE theorem,

$$\begin{aligned} &\langle {}^2\mathrm{F}_{5/2}, M_1 \mid C_q{}^{(2)} \mid {}^2\mathrm{F}_{5/2}, M_1' \rangle \\ &\quad = (-1)^{5/2-M_1} \begin{pmatrix} 5/2 & 2 & 5/2 \\ -M_1 & q & M_1' \end{pmatrix} ({}^2\mathrm{F}_{5/2} \,\|\, C^{(2)} \,\|\, {}^2\mathrm{F}_{5/2}). \end{aligned}$$

The tensor $\mathbf{C}^{(2)}$ acts in the orbital space and not the spin space; from Eq. (1.65),

$$({}^2\mathrm{F}_{5/2} \,\|\, C^{(2)} \,\|\, {}^2\mathrm{F}_{5/2}) = (-1)^{1/2+3+2+5/2} 6 \begin{Bmatrix} 3 & 2 & 3 \\ 5/2 & 1/2 & 5/2 \end{Bmatrix} (\mathrm{f} \,\|\, C^{(2)} \,\|\, \mathrm{f}).$$

The final reduced matrix element is given in Eq. (1.57):

$$(\mathrm{f} \,||\, C^{(2)} \,||\, \mathrm{f}) = -7 \begin{pmatrix} 3 & 2 & 3 \\ 0 & 0 & 0 \end{pmatrix} = -\sqrt{\frac{28}{15}}\,.$$

Substituting back in the previous equation, we find

$$({}^2\mathrm{F}_{5/2} \,||\, C^{(2)} \,||\, {}^2\mathrm{F}_{5/2}) = -12(105)^{-1/2}.$$

It is now a simple matter to evaluate any matrix element of e^2/r_{12} in terms of the product $e^2\langle r^2\rangle^2 r_{AB}^{-5}$. If we limit our attention to first order effects, only those matrix elements in Eq. (5.31) need be calculated for which the energy corresponding to $\langle M_1M_2 |$ is the same as that corresponding to $| M_1'M_2'\rangle$. If, further, we restrict the analysis to those cases where a first order splitting of the quartets and octets occurs (and not merely a displacement of them), the number of matrix elements to be calculated is further reduced. A tabulation of the residue is made in Table 5.1, and the resulting splittings sketched in Fig. 5.4.

TABLE 5.1. Values of $\langle M_1M_2 | e^2/r_{12} | M'_1M'_2\rangle$ in Units of $e^2\langle r^2\rangle^2 r_{AB}^{-5}$

M_1	M_2	M_1'	M_2'	$\langle M_1M_2 \| e^2/r_{12} \| M_1'M_2'\rangle$
5/2	3/2	3/2	5/2	$-\frac{96}{245}$
−5/2	−3/2	−3/2	−5/2	
5/2	1/2	1/2	5/2	$\frac{12}{245}$
−5/2	−1/2	−1/2	−5/2	
3/2	1/2	1/2	3/2	$-\frac{192}{1225}$
−3/2	−1/2	−1/2	−3/2	
3/2	−1/2	−1/2	3/2	$\frac{108}{1225}$
−3/2	1/2	1/2	−3/2	

Although contributions to the splittings are produced by several other effects, such as superexchange and the magnetic-dipole interaction between the spins of the electrons, it is of interest to complete the calculation. According to the Hartree–Fock analysis of Freeman and Watson [84], $\langle r^2\rangle = 1.2a_0^2$. With $r_{AB} = 8.27a_0$ [85], this yields

$$e^2\langle r^2\rangle^2 r_{AB}^{-5} = 8.2\ \mathrm{cm}^{-1}.$$

The splittings should thus be of the order of 2 cm^{-1}. This figure is made unreliable, however, by a number of factors. Chief among these is the polarizability of the intervening charge cloud. But though the absolute values of the splittings of Fig. 5.4 should be treated with extreme caution, the relative magnitudes should be much more reliable. Experimental work has so far been limited to the ground quartet $| \pm 5/2, \pm 5/2\rangle$, so no direct comparison with experiment is possible yet. The interested reader is referred to a review article by Baker [86] for further information.

PROBLEMS

5.1 Use the techniques of Section 5.2 to obtain the single-center expansion

$$r_{12}{}^t\mathbf{C}^{(t)}(\theta_{12}\phi_{12}) = \sum_{k=0}^{t} (-1)^{t-k} r_1{}^k r_2{}^{t-k} \sqrt{\frac{(2t)!}{(2t-2k)!(2k)!}} (\mathbf{C}_1{}^{(k)}\mathbf{C}_2{}^{(t-k)})^{(t)}.$$

5.2 Verify the identity

$$\sum_k \frac{(2a-2k)!(2k-2b)!}{\{(a-k)!(k-b)!\}^2} = 2^{2(a-b)},$$

and use it to show that

$$\sum_k \begin{pmatrix} k & t+2p-k & t \\ 0 & 0 & 0 \end{pmatrix}^2 = 2^{2t} \frac{(2p)!}{(2p+2t+1)!} \left(\frac{(t+p)!}{p!}\right)^2.$$

Use this equation to simplify the recoupling of two separate single-center expansions of $r_{12}{}^{-1}$, thereby obtaining

$$\frac{1}{r_{12}{}^2} = \sum_t \frac{2^{2t}}{r_>{}^2} \frac{(t!)^2}{(2t)!} F\left(\frac{1}{2}, t+1; t+\frac{3}{2}; r_<{}^2/r_>{}^2\right) (r_</r_>)^t (\mathbf{C}_1{}^{(t)} \cdot \mathbf{C}_2{}^{(t)}),$$

where $F(\alpha, \beta; \gamma; z)$ is a hypergeometric function. (The general single-center expansion of $r_{12}{}^n$ has been given by Sack [86a], who used methods analogous to those described in Section 5.1.)

5.3 A sphere of radius a, centered at the origin, comprises a uniform density of electric charge. The electric potential V for $r \geq a$ is $-r^{-1}$. Prove that the potential inside the sphere is given by

$$V = r^2/2a^3 - 3/2a \qquad (a \geq r \geq 0).$$

Prove that the volume integral

$$\int f(r)(\nabla^2 V)\, d\tau$$

tends to $4\pi f(0)$ as $a \to 0$. Deduce that

$$\nabla^2(r^{-1}) = -4\pi\delta(\mathbf{r}).$$

Prove that the vector delta functions $\boldsymbol{\delta}^{(1)}(\mathbf{r})$ of Section 5.3 satisfy

$$\nabla^2(r^{-2}\mathbf{C}^{(1)}) = 4\pi\boldsymbol{\delta}^{(1)}(\mathbf{r}),$$

and generalize these results to define the tensors $\boldsymbol{\delta}^{(k)}(\mathbf{r})$ (see Brink and Satchler [3]).

5.4 Kay *et al.* [70] give the delta function contributions to $r_{12}^{-4}\mathbf{C}^{(3)}(\theta_{12}\phi_{12})$ in the form

$$-\frac{4}{15}\left\{\frac{1}{r_2}\frac{d}{dr_2}[r_2\mathbf{C}^{(3)}(\theta_2\phi_2)] + \frac{1}{2r_2}[\mathbf{C}^{(3)}(\theta_2\phi_2)\mathbf{l}_2^2 - \mathbf{l}_2^2\mathbf{C}^{(3)}(\theta_2\phi_2)]\right\}\delta(\mathbf{r}_1 - \mathbf{r}_2),$$

together with the prescription to integrate first over $\mathbf{r}_1$, and then operate with d/dr_2 and $\mathbf{l}_2^2$. Prove that

$$[\mathbf{C}^{(3)}, \mathbf{l}^2] = -12\mathbf{C}^{(3)} - 4(3)^{1/2}(\mathbf{C}^{(3)}\mathbf{l})^{(3)},$$

and use recoupling techniques to show that the result of Kay *et al.* is equivalent to that given in Section 5.3 by means of Eqs. (5.11)–(5.12).

5.5 Prove that, for a bipolar expansion in region *I* of Fig. 5.3,

$$\begin{aligned} r_{12}^{-j-1}\mathbf{C}_{12}^{(j)} = \sum_{lk} & r_{1A}{}^{l}r_{2B}{}^{k}r_{AB}{}^{-j-k-l-1} \\ & \times \{(2j+2k+2l+1)!/(2j+1)!(2k)!(2l)!\}^{1/2} \\ & \times [(\mathbf{C}_{1A}{}^{(l)}\mathbf{C}_{B2}{}^{(k)})^{(l+k)}\mathbf{C}_{AB}{}^{(l+k+j)}]^{(j)}. \end{aligned}$$

Show that it contains Eq. (5.17) as a special case.

5.6 By examining the integral

$$\int (C_q^{(k)})_1 f(r_1)(\mathbf{C}_1^{(l)}\cdot\mathbf{C}_2^{(l)})\,d\mathbf{r}_1,$$

show that

$$\delta(\mathbf{r}_1 - \mathbf{r}_2) = (4\pi r_1^2)^{-1}\delta(r_1 - r_2)\sum_l (2l+1)(\mathbf{C}_1^{(l)}\cdot\mathbf{C}_2^{(l)}).$$

5.7 In Section 13.31 of his book, Watson [81] gives the integral

$$\int_0^\infty \exp(-p^2t^2)\,J_\nu(at)J_\nu(bt)t\,dt = \frac{1}{2p^2}\exp\left(-\frac{a^2+b^2}{4p^2}\right)I_\nu\left(\frac{ab}{2p^2}\right),$$

where, from his Section 3.71,

$$I_{l+1/2}(z) = (2\pi z)^{-1/2} e^z$$

for large z. Examine the form of the integral as p tends to zero, and hence show that

$$\int_0^\infty J_{l+1/2}(at) J_{l+1/2}(bt) t\, dt = (ab)^{-1/2} \delta(a - b).$$

(This equation is given in Eq. 3.112 of Jackson's book [87].) Use this result in conjunction with the techniques of Section 5.6 to obtain an alternative derivation of the expansion for $\delta(\mathbf{r}_1 - \mathbf{r}_2)$ given in Problem 5.6.

6

Free Diatomic Molecules

6.1 ANGULAR MOMENTUM

Although a number of two-center problems have already been discussed, we have not yet considered the consequences of allowing the two centers to move. If, in addition, a number of circulating electrons are admitted, then we are confronted with a diatomic molecule. To begin with, we suppose that all the particles are spin-free and that we can neglect relativistic effects. Our immediate aim is to construct quantities (such as the angular momenta) that form a useful basis for discussing the properties of diatomic molecules.

The two centers, which constitute the nuclei of the diatomic molecule, are labeled by A and B. Electrons are labeled by t, which we suppose runs from 1 to n. In addition to these labels, we introduce C for the center of mass of the nuclear frame and O for that of the entire molecule. Thus

$$\mathbf{r}_C = (M_A\mathbf{r}_A + M_B\mathbf{r}_B)/(M_A + M_B), \tag{6.1}$$

where M_X is the mass of nucleus X, and

$$\mathbf{r}_O = (M_A\mathbf{r}_A + M_B\mathbf{r}_B + m\sum_{t=1}^{n}\mathbf{r}_t)/(M_A + M_B + nm). \tag{6.2}$$

It is a simple matter to formally write down the total orbital angular momentum $\mathbf{N}$ of the molecule:

$$i\mathbf{N} = \mathbf{r}_A \times \nabla_A + \mathbf{r}_B \times \nabla_B + \sum_{t=1}^{n}\mathbf{r}_t \times \nabla_t. \tag{6.3}$$

We now change variables according to the scheme

$$(\mathbf{r}_A, \mathbf{r}_B, \mathbf{r}_1, \mathbf{r}_2, \ldots, \mathbf{r}_n) \rightarrow (\mathbf{r}_O, \mathbf{r}_{BA}, \mathbf{r}_{1C}, \mathbf{r}_{2C}, \ldots, \mathbf{r}_{nC}), \tag{6.4}$$

where $\mathbf{r}_{XY} = \mathbf{r}_X - \mathbf{r}_Y$. This serves to refer all electronic motion to the center of mass of the nuclear frame.‡ The nuclear motion is represented by the relative coordinates $\mathbf{r}_{AB}$ and the motion of the molecule as a whole. From Eqs. (6.1)–(6.2), and the equation

$$\mathbf{r}_{tC} = \mathbf{r}_t - \mathbf{r}_C = \mathbf{r}_t - (M_A\mathbf{r}_A + M_B\mathbf{r}_B)/(M_A + M_B),$$

we get

$$\nabla_A = \frac{M_A}{M_A + M_B + nm}\nabla_O - \frac{M_A}{M_A + M_B}\sum_{t=1}^{n}\nabla_{tC} - \nabla_{BA},$$

$$\nabla_B = \frac{M_B}{M_A + M_B + nm}\nabla_O - \frac{M_B}{M_A + M_B}\sum_{t=1}^{n}\nabla_{tC} + \nabla_{BA},$$

$$\nabla_t = \frac{m}{M_A + M_B + nm}\nabla_O + \nabla_{tC}. \tag{6.5}$$

The notation is similar to that used in Chapter 5. Thus ∇_{XY} possesses components $\partial/\partial x_{XY}$, $\partial/\partial y_{XY}$, and $\partial/\partial z_{XY}$. It is easy to substitute the expressions for ∇_A, ∇_B, and ∇_t in Eq. (6.3). The result is

$$i\mathbf{N} = \mathbf{r}_O \times \nabla_O + \mathbf{r}_{BA} \times \nabla_{BA} + \sum_{t=1}^{n}\mathbf{r}_{tC} \times \nabla_{tC}. \tag{6.6}$$

The terms on the right correspond to the angular momentum of the molecule as a whole, that of the relative nuclear motion, and the angular mo-

‡ The choice of the center of mass of the nuclear frame as the origin of molecular coordinates is not the only possibility open to us. Indeed, the treatment of HD in Section 9.3 suggests that the geometric center of the nuclei may be more convenient in that case. Options of this kind have been discussed and compared by Pack and Hirschfelder [88].

mentum of the electrons referred to C. The first term may be dropped whenever translational effects are of no interest. This corresponds to taking $\mathbf{p}_O = 0$.

6.2 THE MOLECULAR FRAME

Only comparatively low-lying states of the nuclear motion are of interest to us. The angular momenta of the electrons are thus at least comparable to those of the nuclei. In view of the enormous disparity between m and the nuclear masses M_A and M_B, the average electronic velocity is very much greater than a typical nuclear velocity. This makes it appropriate to refer the motion of the electrons to a frame fixed to the nuclei rather than to an external frame merely centered at C, which is the situation in Eq. (6.6). This approach is one aspect of the Born–Oppenheimer approximation [89]. At this point in the analysis, however, no approximation is made, merely a transformation of variables. The approximation comes later when first order solutions are accepted as adequate.

We identify the external (laboratory) frame with the frame F of Sections 1.3 and 1.6. The coordinates of electron t are (x_{tC}, y_{tC}, z_{tC}) in F. The frame F' is anchored to the internuclear axis. Like F, it has its origin at C. In F', the coordinates of electron t are (ξ_t, η_t, ζ_t). It is convenient to identify the ζ axis with the nuclear axis AB. If its direction runs from A to B (see Fig. 6.1), then the contributions of ζ_t to x_{tC}, y_{tC}, z_{tC} are

$$\zeta_t \sin\theta_{BA} \cos\phi_{BA}, \qquad \zeta_t \sin\theta_{BA} \sin\phi_{BA}, \qquad \zeta_t \cos\theta_{BA}$$

respectively. When these quantities are compared to the inverses of Eqs.

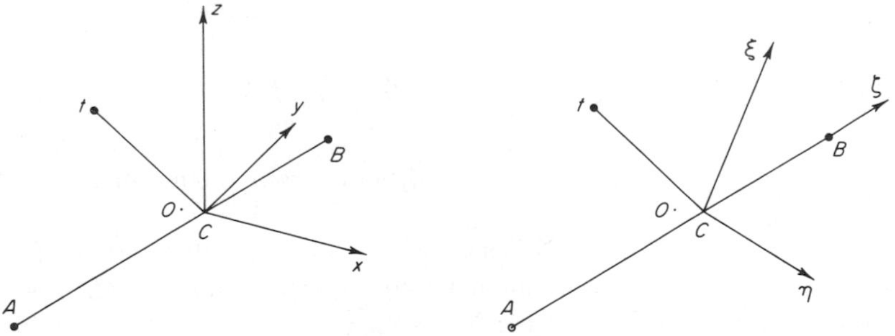

Fig. 6.1. The two frames F and F' superposed on the nuclear axis AB of the diatomic molecule. Both have origins at the center of mass C of the nuclear frame. Points t and O representing an electron and the center of mass of the entire molecule are also indicated.

(1.10), it is at once seen that the identifications

$$\alpha \equiv \phi_{BA}, \qquad \beta \equiv \theta_{BA} \tag{6.7}$$

can be made. The third Euler angle γ cannot be determined; this corresponds to the freedom in fixing the ξ and η axes. We can thus assign any value we wish to γ: it is a parameter at our disposal. For the moment, we shall leave it in the form it is, without specification, although a common choice [90] is $\gamma = \frac{1}{2}\pi$.

The transformation of variables can be represented by the replacement

$$(\mathbf{r}_{1C}, \mathbf{r}_{2C}, \ldots, \mathbf{r}_{nC}, \theta_{BA}, \phi_{BA}) \to (\boldsymbol{\varrho}_1, \boldsymbol{\varrho}_2, \ldots, \boldsymbol{\varrho}_n, \theta, \phi), \tag{6.8}$$

where $\boldsymbol{\varrho}_t$ is the vector with components (ξ_t, η_t, ζ_t), and where $\theta = \theta_{BA}$, and $\phi = \phi_{BA}$. We begin with the term $\mathbf{r}_{tC} \times \boldsymbol{\nabla}_{tC}$ in $i\mathbf{N}$. To express a component of $\boldsymbol{\nabla}_{tC}$ as a linear combination of the operators $\partial/\partial\boldsymbol{\varrho}_t$, the Cartesian tensor $\partial\boldsymbol{\varrho}_t/\partial\mathbf{r}_{tC}$ must be evaluated. It is easy to do this. Since $\mathbf{r}_{tC}$ and (θ_{BA}, ϕ_{BA}) are independent variables, the tensor's nine components are simply the nine functions of α, β, and γ occurring on the right-hand sides of Eqs. (1.10). Owing to the unitarity of this transformation, a component of $\partial/\partial\mathbf{r}_{tC}$ is the same linear combination of the components of $\partial/\partial\boldsymbol{\varrho}_t$ as the corresponding component of $\mathbf{r}_{tC}$ is of the components of $\boldsymbol{\varrho}_t$. In other words, both $\mathbf{r}_{tC}$ and $\partial/\partial\mathbf{r}_{tC}$ transform as vectors. This is not very surprising, of course, but it is always necessary to keep in mind the independence (or otherwise) of the variables being used, and this makes it advisable to proceed cautiously. However, we can now go straight to the result. For $\mathbf{r}_{tC} \times \boldsymbol{\nabla}_{tC}$ must also transform like a vector; and so, from Section 1.7, its components in the laboratory frame F are given by

$$il_{tC}^{(10)} = i(\mathbf{D}^{(11)}(\omega)\,\boldsymbol{\lambda}_t^{(01)})^{(10)}, \tag{6.9}$$

where $\omega \equiv (\phi\theta\gamma)$ and

$$(\boldsymbol{\lambda}_t^{(01)})_\xi = \frac{1}{i}\left(\eta_t \frac{\partial}{\partial\zeta_t} - \zeta_t \frac{\partial}{\partial\eta_t}\right),$$

etc. Equation (6.9) expresses the x, y, and z components of $\mathbf{r}_{tC} \times \boldsymbol{\nabla}_{tC}$ in terms of $(\xi_t\eta_t\zeta_t\theta\phi)$.

The term $\mathbf{r}_{BA} \times \boldsymbol{\nabla}_{BA}$ in $i\mathbf{N}$ is somewhat more difficult to treat. If we put $\mathbf{r}_{BA} \times \boldsymbol{\nabla}_{BA} = il_{BA}$, it can be immediately expressed in terms of $\partial/\partial\theta_{BA}$ and $\partial/\partial\phi_{BA}$ by means of Eqs. (1.6). Now

$$\frac{\partial}{\partial\theta_{BA}} = \frac{\partial\theta}{\partial\theta_{BA}}\frac{\partial}{\partial\theta} + \sum_{t=1}^{n}\left(\frac{\partial\xi_t}{\partial\theta_{BA}}\frac{\partial}{\partial\xi_t} + \frac{\partial\eta_t}{\partial\theta_{BA}}\frac{\partial}{\partial\eta_t} + \frac{\partial\zeta_t}{\partial\theta_{BA}}\frac{\partial}{\partial\zeta_t}\right). \tag{6.10}$$

Although $\theta_{BA} = \theta$, we note, in passing, that $\partial/\partial\theta_{BA} \neq \partial/\partial\theta$. This is because

θ_{BA} and θ possess different independent variables. The partial derivatives in Eq. (6.10) can be evaluated directly. The equation

$$\boldsymbol{\varrho}_t = (\mathbf{D}^{(11)}(\phi_{BA}\theta_{BA}\gamma)\mathbf{r}_{tC}{}^{(10)})^{(01)}$$

is the starting point. However, we can cut short this work by noting that the procedure we are following is equivalent to taking

$$\mathbf{r}^{(01)} = (\mathbf{D}^{(11)}(\alpha\beta\gamma)\mathbf{r}^{(10)})^{(01)}$$

and evaluating $\partial/\partial\beta$ while holding $\mathbf{r}^{(10)}$ constant. The electrons are thus stationary in F; and it is precisely under these conditions that Eqs. (1.23) are valid. If we use those equations to evaluate $\partial/\partial\alpha$ and $\partial/\partial\beta$, we get

$$\begin{aligned}\partial/\partial\alpha &= i\lambda_\xi \sin\beta\cos\gamma - i\lambda_\eta \sin\beta\sin\gamma - i\lambda_\zeta \cos\beta,\\ \partial/\partial\beta &= -i\lambda_\xi \sin\gamma - i\lambda_\eta \cos\gamma.\end{aligned} \tag{6.11}$$

These expressions give the increments to $\partial/\partial\phi_{BA}$ and $\partial/\partial\theta_{BA}$ provided we sum over the electrons. We write

$$\mathbf{L} = \sum_{t=1}^{n} \boldsymbol{\lambda}_t, \tag{6.12}$$

and arrive at the results

$$\begin{aligned}\partial/\partial\phi_{BA} &= \partial/\partial\phi + iL_\xi \sin\theta\cos\gamma - iL_\eta \sin\theta\sin\gamma - iL_\zeta\cos\theta,\\ \partial/\partial\theta_{BA} &= \partial/\partial\theta - iL_\xi\sin\gamma - iL_\eta\cos\gamma.\end{aligned} \tag{6.13}$$

An alternative and more direct method of obtaining these equations (though one not devoid of conceptual problems turning on the independence of variables) is set out in Problem 6.1. It is now straightforward to construct $\boldsymbol{l}_{BA}$ from Eqs. (1.6). The coefficients of L_ξ, L_η, and L_ζ turn out to be very nearly the same as those of ξ, η, and ζ in the inverses to Eqs. (1.10). In fact, we can write

$$\boldsymbol{l}_{BA} = \boldsymbol{l} - (\mathbf{D}^{(11)}(\phi\theta\gamma)\mathbf{L}^{(01)})^{(10)} + \mathbf{T}, \tag{6.14}$$

where the components of $\mathbf{T}$ are given by

$$T_x \pm iT_y = L_\zeta e^{\pm i\phi}\csc\theta, \qquad T_z = 0.$$

Putting the two parts together, we get

$$\begin{aligned}\mathbf{N} &= (\mathbf{D}^{(11)}\mathbf{L}^{(01)})^{(10)} + \boldsymbol{l} - (\mathbf{D}^{(11)}\mathbf{L}^{(01)})^{(10)} + \mathbf{T}\\ &= \boldsymbol{l} + \mathbf{T}\end{aligned} \tag{6.15}$$

for the total orbital angular momentum relative to the center of mass O.

In full detail,

$$N_x \pm iN_y = e^{\pm i\phi}\left(\pm\frac{\partial}{\partial\theta} + i\cot\theta\frac{\partial}{\partial\phi} + L_\zeta \csc\theta\right),$$

$$N_z = \frac{1}{i}\frac{\partial}{\partial\phi}. \tag{6.16}$$

6.3 AN ADDED VARIABLE

Perhaps the most striking feature of Eqs. (6.16) is that, although **N** is the total angular momentum of the molecule (barring spin, of course), the angular momentum of the electrons enters through the single component L_ζ. This component would also not appear explicitly if the nuclear framework possessed some off-axis structure to which the azimuthal angles of the electrons could be referred. The presence of L_ζ in the expressions for $N_\pm$ is something of an embarrassment, since **N** does not commute with the components L_ξ and L_η of **L**. This complicates the separation of nuclear and electronic properties, as well as making it more difficult to apply tensorial methods to the evaluation of matrix elements. The first analysis of this problem appears to have been given by Hougen [91], who solved it by establishing a correspondence between the real physical situation and an artificial one in which the problem does not occur. A somewhat similar approach will be adopted here. Hougen mentions an equivalent way of solving the problem, which he credits to Van Vleck. A fictitious nucleus of mass M_0 is introduced off-axis, thereby defining a direction perpendicular to AB with respect to which the azimuthal angles of the electrons can be defined. Later on, the limit $M_0 \to 0$ can be studied. Such a device removes the inconsistencies from Van Vleck's earlier article [33], where the properties of a diatomic molecule are discussed in terms of quantities defined for the general nonlinear molecule.

The procedure we shall follow is to cast γ in the role of a new variable, subject to the hypothesis that

$$(\varrho_1, \varrho_2, \ldots, \varrho_n, \theta, \phi, \gamma) \tag{6.17}$$

constitutes an independent set. Thus $\partial/\partial\gamma$ commutes with the components L_ξ, L_η, and L_ζ of **L**, just as $\partial/\partial\theta$ and $\partial/\partial\phi$ do. From the point of view of an observer in the laboratory frame F, the effect of $\partial/\partial\gamma$ on a function of the laboratory coordinates $\mathbf{r}_{iC}$ must be reproducible by some linear combination of the differential operators formed from the variables (6.17) *exclud-*

ing γ. Now

$$\begin{aligned}\mathbf{r}_{tC} &= (\mathbf{D}^{(11)}(\phi\theta\gamma)\,\boldsymbol{\varrho}_t^{(01)})^{(10)}\\ &= \sum_q D_{\cdot q}^{(11)}(\phi\theta\gamma)\,(\rho_{0-q}^{(01)})_t(1q,1-q\mid 00).\end{aligned}$$

Since $D_{pq}^{(11)}(\phi\theta\gamma)$ (for any p) depends on γ through $e^{-iq\gamma}$, the effect of $\partial/\partial\gamma$ is to introduce a factor $-iq$ in the sum. On the other hand,

$$[L_\zeta, \rho_{0-q}^{(01)}] = -q\rho_{0-q}^{(01)},$$

and so the equivalence

$$L_\zeta \equiv -i\,\partial/\partial\gamma \tag{6.18}$$

is established. The apparent conflict in sign with the last of Eqs. (1.23) has its origin in the different sets of independent variables in the two cases. There, $\partial/\partial\gamma$ and λ_ζ have compensatory roles to play to ensure that $\mathbf{r}^{(10)}$ remains constant: here, an equivalence is sought under quite different conditions.

The equivalence (6.18) can be expressed by stating that we have two alternative forms for L_ζ: we can take either $-i\,\partial/\partial\gamma$ or

$$\sum_{t=1}^{n}\frac{1}{i}\left(\xi_t\frac{\partial}{\partial\eta_t} - \eta_t\frac{\partial}{\partial\xi_t}\right). \tag{6.19}$$

If we choose the former for the L_ζ that occurs in Eqs. (6.16) and the latter for all other purposes, we can ensure that $\mathbf{L}$ commutes with $\mathbf{N}$, since we have asserted that the set (6.17) comprises independent variables. Equations (6.16) can now be put in the form

$$\begin{aligned}N_x \pm iN_y &= ie^{\pm i\phi}\left(\cot\theta\frac{\partial}{\partial\phi} \mp i\frac{\partial}{\partial\theta} - \frac{1}{\sin\theta}\frac{\partial}{\partial\gamma}\right),\\ N_z &= -i\frac{\partial}{\partial\phi}.\end{aligned} \tag{6.20}$$

These equations parallel Eqs. (1.22) in a very satisfactory way. If we wish to refer $\mathbf{N}$ to the axes of F', we can use Problem 3.2 and Eqs. (1.23) to obtain

$$\begin{aligned}N_\xi \pm iN_\eta &= -ie^{\mp i\gamma}\left(\cot\theta\frac{\partial}{\partial\gamma} \pm i\frac{\partial}{\partial\theta} - \frac{1}{\sin\theta}\frac{\partial}{\partial\phi}\right),\\ N_\zeta &= -i\frac{\partial}{\partial\gamma}.\end{aligned} \tag{6.21}$$

It should be stressed that N_x, N_y, and N_z form the components of an angular momentum vector, but N_ξ, N_η, and N_ζ do not. This point is taken up again in Section 6.4.

Having introduced the variable γ into the analysis, we must expect a new quantum number to be associated with it and thus an increased complexity in the energy spectrum. We must therefore select from the enlarged spectrum that part that corresponds to the physical situation. Fortunately, it is not difficult to do this. The operator L_ζ (given in 6.19) and N_ζ $(= -i\partial/\partial\gamma)$ are functions of independent variables but are equivalent in the laboratory frame F. This equivalence is guaranteed if the bases for describing the diatomic molecule are restricted to those states $|\psi\rangle$ for which

$$(N_\zeta - L_\zeta)\,|\psi\rangle = 0. \tag{6.22}$$

In this way, we avoid the equation $N_\zeta = L_\zeta$, which is inconsistent with $[\mathbf{N}, \mathbf{L}] = 0$, while ensuring the equivalence between N_ζ and L_ζ. The physical content of Eq. (6.22) is simply that the total angular momentum of the molecule about the internuclear axis comes entirely from the electrons.

6.4 GENERATORS OF $R(4)$

Although we have already been using double tensors in our calculations for diatomic molecules, this is probably a good point to reflect a little on their significance. In an equation such as (6.14), the coupled product $(\mathbf{D}^{(11)}\mathbf{L}^{(01)})^{(10)}$ is simply shorthand for three linear combinations of L_ξ, L_η and L_ζ. However, we can evidently pose the problem of finding the angular momentum vectors with respect to which the ranks have the well-defined significance of Eqs. (1.18). Since we have consistently used the first rank of a pair (kk') to denote the tensorial character of an operator in the laboratory frame F, the vector in question here is obviously the total angular momentum $\mathbf{N}^{(10)}$ whose three components N_x, N_y, and N_z are given in Eqs. (6.20).

The transformation of variables has slightly obscured the corresponding solution for k'. We might naively imagine that the required vector is $\mathbf{N}^{(01)}$, but we can see (from Problem 3.2) that the components N_ξ, N_η, and N_ζ do not satisfy the commutation relations of an angular momentum vector; in fact, to match the angular momentum vector $\boldsymbol{\lambda}^{(01)}$ of Eqs. (1.23), we need $-\mathbf{N}^{(01)}$. However, this is not sufficient in itself, since $\mathbf{L}^{(01)}$ commutes with $-\mathbf{N}^{(01)}$ and so belies its rank of 1 in behaving like a scalar. The combination

$$\mathbf{G}^{(01)} = \mathbf{L}^{(01)} - \mathbf{N}^{(01)} \tag{6.23}$$

is required. Being composed of the sum of two commuting angular momenta ($\mathbf{L}^{(01)}$ and $-\mathbf{N}^{(01)}$), $\mathbf{G}^{(01)}$ is itself an angular momentum vector. Moreover, $\mathbf{L}^{(01)}$ tests functions of the $\boldsymbol{\varrho}_t$'s for their tensorial character, while $-\mathbf{N}^{(01)}$ plays a similar role for functions of $(\phi\theta\gamma)$. It is readily seen that

$$[\mathbf{N}^{(10)}, \mathbf{G}^{(01)}] = -[\mathbf{N}^{(10)}, \mathbf{N}^{(01)}] = 0, \tag{6.24}$$

so $\mathbf{N}^{(10)}$ and $\mathbf{G}^{(01)}$ can be taken as the generators of an $R(4)$ algebra. The ranks (kk') define representations in terms of the $R(3) \times R(3)$ structure of $R(4)$.

An interesting aspect of this analysis is that Eq. (6.22) appears as

$$G_\zeta \mid \psi \rangle = 0.$$

Thus the only acceptable states $\mid \psi \rangle$ of a diatomic molecule are those for which the eigenvalue Γ of G_ζ is zero. This result, like others of this section, have been derived under the assumption that we can neglect the spins of the particles. The modifications that must be made when this condition is relaxed are given in Problem 6.6.

For future use, we define

$$\mathbf{R} = -\mathbf{G} = \mathbf{N} - \mathbf{L}. \tag{6.25}$$

By examining the commutators $[R_{p0}{}^{(10)}, R_{q0}{}^{(10)}]$, we can show that $\mathbf{R}^{(10)}$ is an angular momentum vector. A more direct way is to use Eqs. (6.14)–(6.15) to convert $\mathbf{R}$ to the intermediate coordinate system in which the angles θ_{BA} and ϕ_{BA} appear. We at once get

$$\mathbf{R} = \mathbf{l}_{BA}. \tag{6.26}$$

Thus $\mathbf{R}$ is the angular momentum coming from the nuclear motion. It should be borne in mind, however, that our choice of variables for expressing $\mathbf{N}$ and $\mathbf{L}$ makes it a function of the electrons' coordinates as well as those of the nuclei.

6.5 THE TOTAL SPIN-INDEPENDENT HAMILTONIAN

It is straightforward to write down formal expressions for the potential and kinetic energies of the particles comprising the diatomic molecule. If we continue to neglect spin, the total kinetic energy T is given by

$$T = -\frac{1}{2}\hbar^2 \left(\frac{1}{M_A} \nabla_A{}^2 + \frac{1}{M_B} \nabla_B{}^2 + \frac{1}{m} \sum_{t=1}^{n} \nabla_t{}^2 \right). \tag{6.27}$$

With the aid of Eqs. (6.5), this becomes

$$T = -\frac{1}{2}\hbar^2\left(\frac{1}{M_A + M_B + nm}\nabla_O^2 + \frac{1}{\mu}\nabla_{BA}^2 + \frac{1}{\mu'}\sum_{t=1}^{n}\nabla_{tC}^2\right.$$

$$\left.+\frac{1}{M_A + M_B}\sum_{t\neq v}\nabla_{tC}\cdot\nabla_{vC}\right), \tag{6.28}$$

where the molecular reduced mass μ is given by

$$\mu = M_A M_B/(M_A + M_B), \tag{6.29}$$

and the electronic reduced mass μ' by

$$\mu' = m(M_A + M_B)/(M_A + M_B + m). \tag{6.30}$$

Proceeding as before, the transformation represented schematically by (6.8) is carried out. The vector operator ∇_{tC} is now formed from the components $\partial/\partial\xi_t$, $\partial/\partial\eta_t$, and $\partial/\partial\zeta_t$. The term in ∇_{BA}^2 can be simplified by using Eq. (2.17). Writing $r_{BA} \equiv r$, we find

$$\nabla_{BA}^2 = \frac{1}{r^2}\frac{\partial}{\partial r}\left(r^2\frac{\partial}{\partial r}\right) - \frac{1}{r^2}\mathbf{l}_{BA}^2,$$

where, from Eqs. (6.13)–(6.14) and (6.25)–(6.26),

$$\mathbf{l}_{BA}^2 = (\mathbf{N} - \mathbf{L})^2 = \mathbf{R}^2.$$

To construct the total spin-independent Hamiltonian H, we add two terms to T. The first, $U(\boldsymbol{\rho}, r)$, represents the potential energy of the electrons in their own field and that of the nuclei. The second is written $V(r)$ $(= Z_A Z_B e^2/r)$, and corresponds to the electrostatic repulsion of the two nuclei. In anticipation of the subsequent analysis, we break H up into four parts, corresponding to the electronic, vibrational, rotational, and translational motions. Thus

$$H = H_e + H_v + H_r + H_t,$$

where

$$H_e = -\frac{\hbar^2}{2\mu'}\sum_{t=1}^{n}\nabla_{tC}^2 - \frac{\hbar^2}{2(M_A + M_B)}\sum_{t\neq v}\nabla_{tC}\cdot\nabla_{vC} + U(\boldsymbol{\rho}, \mathbf{r}),$$

$$H_v = -\frac{\hbar^2}{2\mu r^2}\frac{\partial}{\partial r}\left(r^2\frac{\partial}{\partial r}\right) + V(r),$$

$$H_r = \frac{\hbar^2}{2\mu r^2}(\mathbf{N} - \mathbf{L})^2,$$

$$H_t = -\frac{\hbar^2}{2(M_A + M_B + mn)}\nabla_O^2. \tag{6.31}$$

This separation of H corresponds roughly to the relative magnitudes of the different parts, and prepares the ground for the application of perturbation theory. A formal justification for the separation has been provided by Born and Oppenheimer [89], who use $(m/M)^{1/4}$ as an expansion parameter, where M is a typical nuclear mass (see also Born and Huang [92]). In their treatment, the electronic motion (with fixed nuclei) appears in zeroth order, the harmonic motion of the nuclei in second order, and the rotations and translations in fourth order. Coupling between the different motions appears in fifth and higher orders. This approach has the curious result of postponing the comparatively trivial analysis of the translations and rotations until rather high orders of perturbation theory are reached. A more direct procedure is to separate off the translations immediately, then pass to H_e and H_r before treating H_v. In this way, the effect of rotation is automatically incorporated into the vibrational spectrum. The precise way to break up H, and the order in which the parts are to be considered, entail decisions of the kind that occur whenever a complicated Hamiltonian is treated by perturbation theory.

6.6 THE PRIMITIVE ENERGY SPECTRUM

The electronic Hamiltonian H_e represents the motion of the n electrons moving in the electrostatic field produced by themselves and the two fixed nuclei. The fact that the nuclear masses are not infinite gives rise to the term involving $\nabla_{tC} \cdot \nabla_{vC}$; in the atomic case (corresponding to $r = 0$), this term is the source of the so-called specific isotope effect. Even without this comparatively minor term, the problem of finding the eigenvalues E_e of H_e is, of course, a major area of research in its own right. Only a few general features are of immediate interest to us here. First, we note that E_e is a function of the internuclear separation r. Secondly, H_e is invariant under rotations about the nuclear axis AB. That is, $[L_\zeta, H_e] = 0$. Thus the various eigenfunctions of H_e, which, in the ordinary way, we might distinguish simply by some running index ν, can be further classified by the eigenvalue Λ of L_ζ. We can therefore write

$$H_e \mid \nu\Lambda\rangle = E_{e\nu}(r) \mid \nu\Lambda\rangle.$$

The eigenfunctions $\mid \nu\Lambda\rangle$, like $E_{e\nu}$, depend on r.

The existence of the quantum number Λ immediately leads to selection rules on L_ξ and L_η. For

$$\begin{aligned} L_\zeta\{(L_\xi \pm iL_\eta) \mid \nu\Lambda\rangle\} &= (L_\xi \pm iL_\eta)(L_\zeta \pm 1) \mid \nu\Lambda\rangle \\ &= (\Lambda \pm 1)\{(L_\xi \pm iL_\eta) \mid \nu\Lambda\rangle\}, \end{aligned}$$

and we have only to set the bra $\langle \nu'\Lambda' |$ on the left of these expressions to get

$$\langle \nu'\Lambda' | L_\xi \pm iL_\eta | \nu\Lambda\rangle(\Lambda' - \Lambda \mp 1) = 0. \tag{6.32}$$

Thus L_ξ and L_η can only connect electronic states differing by unity in Λ.

We next consider the rotational Hamiltonian, H_r. Rather surprisingly, the expression for H_r given in Eqs. (6.31) exhibits spherical symmetry. This is because the condition that the eigenvalues of L_ζ and N_ζ be equal has not yet been applied. It is worth noticing that the moments of inertia I_1 and I_2 of the nuclear frame about the ξ and η axes are both μr^2, so the part of H_r involving $N_\xi^2 + N_\eta^2$ is identical to the first two terms in the expression for the kinetic energy of a rigid rotator as given in Eq. (3.3). Of those terms in H_r that depend on the Euler angles, we can, for the moment, set aside $L_\xi N_\xi + L_\eta N_\eta$, since this operator can only connect different electronic states, and we are assuming that H_e is much more important than H_r. We are left with a term proportional to $\mathbf{N}^2 - 2L_\zeta N_\zeta$. For a given electronic state, L_ζ possesses the eigenvalue Λ. The eigenstates of $\mathbf{N}^2 - 2\Lambda N_\zeta$ are functions of the Euler angles, but there is no need to formally write down an expansion in terms of the bases $| NM -\Lambda'\rangle$ of Eq. (3.6), since each one, in itself, is an eigenstate:

$$(\mathbf{N}^2 - 2\Lambda N_\zeta) | NM -\Lambda'\rangle = \{N(N+1) - 2\Lambda\Lambda'\} | NM -\Lambda'\rangle. \tag{6.33}$$

To get this result, we have used the fact that $-\Lambda'$ is the eigenvalue of $-N_\zeta$. The condition (6.22) now gives $\Lambda = \Lambda'$. The electronic and rotational kets can be written in the combined form

$$| [\nu\Lambda][NM -\Lambda]\rangle.$$

It is to be noted, however, that these kets are eigenstates of only a part of H_r. The residue is carried forward into H_v.

The variable r is the only one left to examine. Its associated quantum number (say v) is to be used in describing the eigenstates of the vibrational Hamiltonian H_v. The totality of all possible kets

$$| [\nu\Lambda][v][NM -\Lambda]\rangle$$

constitute a complete set with respect to which the total Hamiltonian H is ideally evaluated. (In principle, electronic states in the continuum are included.) We cannot immediately treat the vibrational equation because the eigenvalues of H_e and H_r, as well as the electronic eigenfunctions $| \nu\Lambda\rangle$, are themselves functions of r. A common approach, called the Born Adiabatic Approximation, is to limit attention to the diagonal matrix elements

$$E_0 = \langle [\nu\Lambda][v][NM -\Lambda] | H | [\nu\Lambda][v][NM -\Lambda]\rangle, \tag{6.34}$$

and to take these as constituting the approximate eigenvalues of the total Hamiltonian. The approximation thus consists in not going beyond the lowest order of perturbation theory. The parts H_e and H_r of H can be evaluated to give

$$\begin{aligned} E_0 = \langle [\nu\Lambda][v][NM-\Lambda] \mid & E_{e\nu}(r) + V(r) \\ & + (\hbar^2/2\mu r^2)\{N(N+1) - \Lambda^2 + L_\xi^2 + L_\eta^2\} \\ & - (\hbar^2/2\mu r^2)(\partial/\partial r)(r^2\,\partial/\partial r) \mid [\nu\Lambda][v][NM-\Lambda]\rangle. \end{aligned}$$

We can now reduce the number of variables by integrating over γ, θ, and ϕ; the labels $[NM-\Lambda]$ disappear from the kets. The term involving the derivatives with respect to r can be simplified by using the fact that an eigenfunction is transformed into its complex conjugate under time reversal (see Problem 6.5). This operation in no way affects $\partial/\partial r$; hence

$$\begin{aligned} \langle \nu\Lambda \mid (\partial/\partial r) \mid \nu\Lambda\rangle &= \langle \nu\Lambda \mid (\partial/\partial r) \mid \nu\Lambda\rangle^* \\ &= \tfrac{1}{2}(\partial/\partial r)\langle \nu\Lambda \mid \nu\Lambda\rangle = 0. \end{aligned}$$

Integration over the electronic coordinates, where possible, now yields

$$E_0 = \langle v \mid Z(r) \mid v\rangle,$$

where

$$\begin{aligned} Z(r) = {} & E_{e\nu}(r) + V(r) + (\hbar^2/2\mu r^2)\{N(N+1) - \Lambda^2 \\ & + \langle \nu\Lambda \mid L_\xi^2 + L_\eta^2 \mid \nu\Lambda\rangle\} \\ & - (\hbar^2/2\mu r^2)(\partial/\partial r)(r^2\partial/\partial r) - (\hbar^2/2\mu)\langle \nu\Lambda \mid (\partial^2/\partial r^2) \mid \nu\Lambda\rangle. \end{aligned} \tag{6.35}$$

We have not as yet specified $\mid v\rangle$, which may be any complete set of functions of a single variable. However, the basis will most nearly approximate to the actual physical situation if $Z(r)$ is diagonal with respect to v. That is, the equation

$$\left[\frac{\hbar^2}{2\mu r^2}\frac{\partial}{\partial r}\left(r^2\frac{\partial}{\partial r}\right) - W(r) - \frac{\hbar^2}{2\mu r^2}N(N+1) + E_0\right] \mid v\rangle = 0 \tag{6.36}$$

should be valid, where the potential function $W(r)$ is given by

$$\begin{aligned} W(r) = {} & E_{e\nu}(r) + V(r) + (\hbar^2/2\mu)\langle \nu\Lambda \mid (L_\xi^2 + L_\eta^2)r^{-2} - (\partial^2/\partial r^2) \mid \nu\Lambda\rangle \\ & - (\hbar^2/2\mu r^2)\Lambda^2. \end{aligned} \tag{6.37}$$

Not surprisingly, the potential $W(r)$ depends on the electronic structure of the molecule. Different electronic states thus lead to different sets of vibrational functions $\mid v\rangle$. The actual calculation of $W(r)$ is, of course, a problem of great complexity. For the purpose of fitting experimental data, it is convenient to parametrize it in some way. Several of the possibilities open to us are discussed in Section 6.6. To complete the present discussion of the Hamiltonian, we make the simplest conceivable assumptions: namely, that the rotational term involving $N(N+1)$ is sufficiently small to be momentarily set aside, and that $W(r)$ can be approximated by the parabolic form

$$W(r) = W(r_e) + \tfrac{1}{2}k^2(r - r_e)^2, \qquad (6.38)$$

where r_e is the equilibrium separation of the nuclei. Equation (6.36) now reduces to that of a simple-harmonic oscillator. The eigenvalues E_0' are given by

$$E_0' = W(r_e) + (v + \tfrac{1}{2})\hbar\omega \qquad (v = 0, 1, 2, \ldots),$$

where $\omega = k\mu^{-1/2}$. The kets $|\,v\rangle$ can now be assigned the explicit form

$$|\,v\rangle = \frac{1}{r}\left(\frac{\kappa}{\pi}\right)^{1/4}\left(\frac{1}{2^v v!}\right)^{1/2} \exp\left(-\frac{1}{2}\kappa(r - r_e)^2\right)H_v\{(r - r_e)(\kappa)^{1/2}\}, \qquad (6.39)$$

where $\kappa = \mu\omega/\hbar$. The Hermite polynomial $H_v(\sigma)$ is given by

$$H_v(\sigma) = (-1)^v e^{\sigma^2} \frac{d^v}{d\sigma^v} e^{-\sigma^2}. \qquad (6.40)$$

In the limit of large κ (for which the amplitude of the vibrations is small compared to r_e),

$$\langle v\,|\,r^{-2}\,|\,v\rangle = r_e^{-2}.$$

We can therefore write, in the limit of small vibrations,

$$\langle v\,|\,(\hbar^2/2\mu r^2)N(N+1)\,|\,v\rangle = (\hbar^2/2\mu r_e^2)N(N+1)$$

for the rotational term of Eq. (6.36). Thus, under these various approximations and assumptions,

$$E_0 = W(r_e) + (v + \tfrac{1}{2})\hbar\omega + (\hbar^2/2\mu r_e^2)N(N+1). \qquad (6.41)$$

This simple expression, which clearly distinguishes the electronic, vibrational, and rotational contributions to E_0, has been obtained at the expense of neglecting all couplings between these various parts. However, in the process we have obtained an explicit form for the kets $|\,[\nu\Lambda][v][NM\ -\Lambda]\rangle$, and these can now be used to examine and evaluate the couplings in question. It is one of the great strengths of perturbation theory that neglected contributions to the Hamiltonian can, in principle, be treated to any desired accuracy once the zeroth order eigenfunctions are known.

6.7 THE ROTATING OSCILLATOR AND THE HYDROGEN ATOM

Equation (6.36) is formally the same as that of an electron with angular momentum quantum number N moving in a central potential defined by $W(r)$. In the case of a single-electron atom with nuclear charge Ze, we have $W(r) = -Ze^2/r$. This function possesses a positive slope everywhere, and it is only the counteracting centrifugal term $\hbar^2 N(N+1)/2\mu r^2$ that makes it possible for a potential well to be formed. For a molecule, $W(r)$ itself possesses a minimum; the centrifugal term is much less important in determining the equilibrium separation. It is possible to parametrize $W(r)$ so that a continuous transition can be made from the rotating oscillator to the hydrogen atom [93]. If we write

$$| v\rangle = r^{-1} | v), \qquad 2\mu E_0/\hbar^2 = \epsilon_0,$$

$$W(r) = (\hbar^2/2\mu)(a^2/r^2 - 2b/r), \tag{6.42}$$

then Eq. (6.36) becomes

$$\left[\frac{\partial^2}{\partial r^2} - \frac{N(N+1)+a^2}{r^2} + \frac{2b}{r} + \epsilon_0\right] | v) = 0. \tag{6.43}$$

For $a = 0$, the hydrogenic equation is obtained. As a increases, the relative importance of the centrifugal term diminishes, and the rotating oscillator is obtained in the limit $a \to \infty$.

It is of interest to find the eigenvalues ϵ_0 for an arbitrary value of a. Since Eq. (6.43) has the form of a hydrogenic equation characterized by an effective angular momentum quantum number l given by

$$N(N+1) + a^2 = l(l+1), \tag{6.44}$$

the problem presents no special difficulties other than those arising from the possibility that l is not integral. We make the usual substitutions for $| v)$ to obtain a power-series expansion for which the ratio between any pair of adjacent terms is known. The demand that the series terminates yields the eigenvalues

$$\epsilon_0 = -b^2/(v + l + 1)^2, \tag{6.45}$$

where l is the positive solution of Eq. (6.44) and $v = 0, 1, 2, \ldots$. The (unnormalized) eigenfunctions are

$$| v\rangle = r^{-1} | v) = e^{-br/(v+l+1)} r^l \Phi[-v, 2l+2, 2br/(v+l+1)],$$

where Φ is a confluent hypergeometric function [67]. For $a = 0$, we have $l = N$; the usual hydrogenic spectrum is at once obtained by writing $v = n - l - 1$, and Φ becomes a Laguerre polynomial.

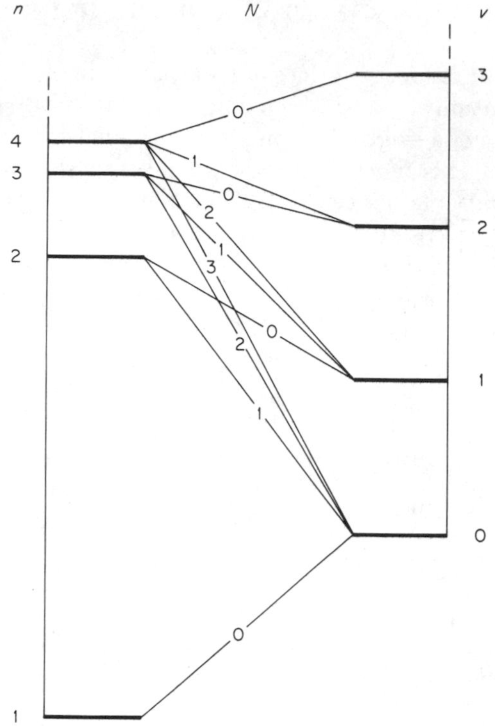

Fig. 6.2. Connectivity between the hydrogenic states (on the left) and those of the rotating oscillator (on the right). Only the four lowest levels of both structures are shown. Each oscillator level v is the junction of an infinity of lines corresponding to the angular momenta $N = 0, 1, 2, \ldots$ that originate from the hydrogenic levels for which $n = v + 1, v + 2, \ldots$. Only connections and not actual energies are shown in the intermediate region: the latter may be obtained from Eq. (6.45).

The situation for large a is rather more interesting. Expanding ϵ_0 in this limit, we get

$$\begin{aligned}\epsilon_0/b^2 = {} & -a^{-2} + 2a^{-3}(v + \tfrac{1}{2}) - 3a^{-4}(v + \tfrac{1}{2})^2 + \cdots \\ & + a^{-4}N(N + 1) - a^{-6}N^2(N + 1)^2 + \cdots, \end{aligned} \tag{6.46}$$

as well as cross terms involving both vibrational and rotational quantum numbers. Equation (6.46) shows that both the vibrational and rotational spectra tend, in the first instance, to contract as higher quantum numbers are reached. The connectivity between the levels of the hydrogen atom and those of the rotating oscillator are sketched in Fig. 6.2.

The expression we have been using for $W(r)$ involves only two parameters, a and b. Better fits with experiment are obtained if we use the three-parameter Morse potential, defined by [94]

$$W(r) = W(r_e) + D(1 - \exp(-a(r - r_e)))^2. \qquad (6.47)$$

Like the $W(r)$ of Eq. (6.42), it provides a well that deviates from the perfect parabolic shape, and hence gives rise to anharmonic terms in the energy spectrum (see Problem 6.4). A more elaborate potential has been studied by Dunham [95] and further discussed by Sandeman [96].

6.8 SPIN

The existence of electron spin, in itself, entails no substantial change in the preceding analysis. Of course, the role that it plays in the Pauli exclusion principle determines the acceptable orbital eigenfunctions of the electrons; but until interaction terms involving spin appear in the Hamiltonian, the only modification we need to make is to include spin eigenstates at appropriate points throughout the analysis. The molecule rotates and vibrates while the spins preserve their orientations in the laboratory frame. The total spin **S** of the electrons commutes with H, and eigenstates of the type

$$|\,[\nu\Lambda][v][NM\ -\Lambda][SM_S]\rangle \qquad (6.48)$$

are required. Since **N** also commutes with H, the eigenvalues of H are degenerate with respect to M and M_S. Both these quantum numbers refer to the z axis of the laboratory, and we may readily form those linear combinations of the kets (6.48) that couple **N** and **S** to a resultant **J**. In a compressed notation, these new kets may be written $|\,(NS)JM_J\rangle$. They are rather more convenient than the uncoupled ones (6.48), since, as soon as any interaction within the molecule involving spins is introduced, the total angular momentum quantum number J is preserved as a *good* quantum number. The new kets are thus ideal basis functions provided the interaction is small. The kind of coupling for which the kets $|\,(NS)J\rangle$ are good first approximations is called Hund's case (b) [97].

It may happen, on the other hand, that the interactions involving electron spin and the particles making up the diatomic molecule are strong. We might consider coupling **S** with **L** before projecting the resultant on the internuclear axis (Hund's case (c)). More frequently, it is better to directly quantize **S** along the axis; the component S_ζ is thus assigned a quantum number Σ analogous to the eigenvalue Λ of L_ζ. This is Hund's case (a). The spin eigenfunctions are now referred to the molecular frame, and this imposes a number of changes in the analysis of Sections 6.2–6.7. The principal complication arises from the fact that **S**, unlike **L**, does not have a simple representation in terms of the electrons' coordinates. Fortunately, the method of Problem 6.1 circumvents this difficulty. If we wish to refer the electrons' spins, as well as their orbital motion, to the molecular frame F', we have only to use expressions for $D(\omega)$ in which $\mathbf{L}$ is replaced by $\mathbf{P}$, where

$$\mathbf{P} = \mathbf{S} + \mathbf{L}. \tag{6.49}$$

It is a trivial matter to repeat the analysis of Problem 6.1 for $\mathbf{P}$ instead of $\mathbf{L}$. The result is that Eqs. (6.13) remain valid provided L_ξ, L_η, and L_ζ are replaced by P_ξ, P_η, and P_ζ. In the process, the interpretation of $\partial/\partial\theta$ and $\partial/\partial\phi$ undergoes a subtle change: these operators now imply differentiation in which not only the electrons' coordinates $\boldsymbol{\varrho}_i$ are held fixed, but also their spin orientations in F'. The total angular momentum of the molecule—now called $\mathbf{J}$—possesses components that are the same functions of $\partial/\partial\phi$, $\partial/\partial\theta$, and $\partial/\partial\gamma$ as the components of $\mathbf{N}$ were. That is, $\mathbf{J}$ is given by the right-hand sides of Eqs. (6.20)–(6.21). Equation (6.22) is replaced by

$$(J_\zeta - P_\zeta) \mid \psi\rangle = 0, \tag{6.50}$$

and $\mathbf{J}$ commutes with $\mathbf{P}$. The rotational Hamiltonian H_r of Eqs. (6.31) becomes

$$H_\mathrm{r}' = (\hbar^2/2\mu r^2)(\mathbf{J} - \mathbf{P})^2, \tag{6.51}$$

and the analysis of Section 6.6 goes through as before except for some notational adjustments. Thus $J(J+1)$ replaces $N(N+1)$ in Eq. (6.36). The new kets are written

$$\mid [\nu\Lambda S\Sigma][v][JM_J\ -\Omega]\rangle, \tag{6.52}$$

where $\Omega = \Lambda + \Sigma$. This last equation follows from an application of Eq. (6.50).‡

In deriving (6.52), no actual term involving spin has been added to the Hamiltonian. We have merely described an alternative basis. But unless an appropriate interaction involving spin exists, the kets (6.52) will provide a very poor approximation to the eigenfunction. Of course, any ket exhibiting a coupling of Hund's case (a) can be expressed as a linear superposition of kets corresponding to Hund's case (b) and vice-versa. (See Problem 9.1.)

‡ Our symbols Λ, Σ, and Ω are not quite identical to those recommended by the 1953 Joint Commission for Spectroscopy [98]. The latter (denoted by primes) are related to the former as follows:

$$\Lambda' = |\Lambda|, \qquad \Sigma' = \Sigma, \qquad \Omega' = |\Lambda' + \Sigma'| = ||\Lambda| + \Sigma|.$$

The absence of an explicit coordinate representation for the eigenfunctions of spin makes it a little difficult to determine how they transform under the improper rotations of inversion or reflection. All we can do is to write down the equations satisfied by the two basic spin eigenfunctions α and β (corresponding to $m_s = +\frac{1}{2}$ and $-\frac{1}{2}$), and examine how each component part behaves under the improper rotation. A crucial assumption—which can be justified by Dirac's theory of the electron—is that **s** transforms in an identical way to ***l***. In the case of the inversion operation I, we know ***l*** is invariant, and so **s** must be too. The equations

$$s_z\alpha = \tfrac{1}{2}\alpha, \qquad s_z\beta = -\tfrac{1}{2}\beta$$

thus become

$$s_z(I\alpha) = \tfrac{1}{2}(I\alpha), \qquad s_z(I\beta) = -\tfrac{1}{2}(I\beta),$$

from which we can deduce that

$$I\alpha = \alpha, \qquad I\beta = \beta. \tag{6.53}$$

Although it might appear that two phase factors have been introduced here, the necessity of preserving the equations

$$s_-\alpha = \beta, \qquad s_+\beta = \alpha$$

under inversion limits our freedom to a single arbitrary phase. Of course, the form of Eqs. (6.53) is particularly convenient, since the parity of many-electron eigenfunctions can now be analyzed without having to pay any attention to spin at all.

A reflection R can be constructed by following the inversion I by a rotation by π. Since I is ineffective for spin eigenfunctions, we can, in this case, pass immediately to the rotation. For example, a reflection in the xz plane can be performed by an inversion plus a rotation by π about the y axis. From Section 1.4, we find

$$\begin{aligned} D(0\pi0)\alpha &= \mathfrak{D}_{\frac{1}{2}\frac{1}{2}}^{\frac{1}{2}}(0\pi0)\alpha + \mathfrak{D}_{-\frac{1}{2}\frac{1}{2}}^{\frac{1}{2}}(0\pi0)\beta \\ &= \beta, \\ D(0\pi0)\beta &= -\alpha. \end{aligned}$$

We can conclude that

$$R\alpha = \beta, \qquad R\beta = -\alpha. \tag{6.54}$$

Our use of π rather than $-\pi$ for the angle of rotation may seem innocuous enough, but these two rotations are not equivalent for eigenfunctions characterized by half-integral angular momenta. So Eqs. (6.54) conceal another phase choice.

6.9 ELECTRONIC STRUCTURE

The purely electronic parts $|\nu\Lambda SM_S\rangle$ or $|\nu\Lambda S\Sigma\rangle$ of the total molecular state are functions of the spatial and spin coordinates of the n electrons making up the diatomic molecule. As in the atomic case, the electrons are assigned to orbitals which, in principle, can be obtained by a self-consistent procedure. Since the electronic Hamiltonian H_e is invariant under rotations about the ζ axis, it is highly convenient to take the molecular orbitals ψ to be eigenfunctions of λ_ζ, the single-electron contribution to L_ζ. This means that they should have an azimuthal dependence of the type $e^{im\varphi}$ (where $\tan\varphi = \eta/\xi$ and m is the eigenvalue of λ_ζ), just as the H_2^+ states of Eq. (4.11) do. To these orbitals are attached the spin eigenfunctions α or β, corresponding to $m_s = \pm\frac{1}{2}$ (referred to the z axis for Hund's case (b) and to the ζ axis for Hund's case (a)). The Pauli exclusion principle is satisfied by forming determinantal product states, and $|\nu\Lambda SM_S\rangle$ or $|\nu\Lambda S\Sigma\rangle$ is expressed as some linear combination of these determinants.

The electronic Hamiltonian H_e is invariant under the reflection R in the plane containing the internuclear axis. This remains true when terms involving spin are included in H_e. The operator R, which is sometimes called the Kronig reflection, can thus be used to label the electronic eigenfunctions. If we take the $\xi\zeta$ plane as the reflection plane, the angle φ_t for electron t becomes $2\pi - \varphi_t$. Since m is integral, the effect of R on every exponential $e^{im\varphi}$ is equivalent to reversing the sign of m. It looks, then, as if the action of R on an orbital ψ_m is to convert it to the orbital ψ_{-m}. It is important to recognize, however, that the relative phases of ψ_m and ψ_{-m} are still at our disposal, and we could perfectly well *define* ψ_{-m} as $-R\psi_m$, for example. We can get some guidance by looking at the atomic case. There, we find

$$RY_{lm}(\vartheta, \varphi) = Y_{lm}(\vartheta, -\varphi) = Y_{lm}^*(\vartheta, \varphi) = (-1)^m Y_{l-m}(\vartheta, \varphi).$$

Thus a reflection introduces the phase $(-1)^m$ as well as reversing the sign of m. It is convenient to define the relative phases of ψ_m and ψ_{-m} such that

$$R\psi_m = (-1)^m \psi_{-m}, \tag{6.55}$$

for then we can easily pass to atomic eigenfunctions in the united-atom limit. In the event that l (the quantum number for $\boldsymbol{\lambda}^2$) is a good quantum number, we can also transform one orbital into another by means of the shift operators $\lambda_\xi \pm i\lambda_\eta$ without having to concern ourselves with questions of relative phase.

Turning now to the states $|\nu\Lambda SM_S\rangle$ (for Hund's case (b)), we first note that the spin eigenfunctions are defined in the frame F of the laboratory and can thus be regarded as untouched by the reflection R. Since L_ζ is the sum of the operators $(\lambda_\zeta)_t$ for the n electrons of the molecule, the sum of the values of m appearing in any determinant contributing to $|\nu\Lambda SM_S\rangle$ must be Λ. We therefore expect R to reverse the sign of Λ. The association of Λ with $-\Lambda$ makes it convenient to use the labels Σ, Π, Δ, Φ, ... for states corresponding to $|\Lambda| = 0, 1, 2, 3, \ldots$. The notation parallels the atomic sequence S, P, D, F, ... for $L = 0, 1, 2, 3, \ldots$. The analogy is sustained by using the lowercase letters σ, π, δ, ϕ, ... for single-electron states.

If $\Lambda = 0$, the reflection operator R can only produce a multiple of the original state when it acts on $|\nu\Lambda SM_S\rangle$. More precisely, it can only produce the state itself or its negative, since two successive replacements of the type $\varphi \to -\varphi$ are equivalent to no change at all: whence $R^2 \equiv 1$ for orbitals. The positive or negative sign is attached to a Σ state by writing Σ^+ or Σ^-.

If, on the other hand, $\Lambda \neq 0$, the question arises as to the definition of $|\nu -\Lambda\, SM_S\rangle$. A very simple and obvious way is to take the linear combination of determinants that represents $|\nu\Lambda SM_S\rangle$ and replace every orbital ψ_m by ψ_{-m}. Then we have, from Eq. (6.55),

$$R\,|\nu\Lambda SM_S\rangle = (-1)^{\Sigma m}\,|\nu -\Lambda\, SM_S\rangle = (-1)^{\Lambda}\,|\nu -\Lambda\, SM_S\rangle, \quad (6.56)$$

provided $\Lambda \neq 0$. This is the convention we shall adopt here. However, it should be noticed that an alternative definition of $|\nu -\Lambda SM_S\rangle$ would be possible if L were a good quantum number, as in the atomic case. We could then write (for positive Λ)

$$|\nu L -\Lambda SM_S\rangle = \{(L-\Lambda)!/(L+\Lambda)!\}L_-^{2\Lambda}\,|\nu L\Lambda SM_S\rangle.$$

This second definition rests on the phase conventions of angular momentum theory, and we cannot immediately tell whether there is a conflict with the first definition. To examine this point, we imagine the state $|\nu L\Lambda\rangle$ to be formed from a superposition of coupled states of the type

$$|(L_{n-1}l_n)L\Lambda\rangle, \quad (6.57)$$

where L_{n-1} is, at the same time, regarded as a superposition of the coupled states

$$|(L_{n-2}l_{n-1})L_{n-1}\rangle,$$

and so on. The symbol l_t stands for the azimuthal quantum number of electron t. The uncoupled component

$$|(l_1m_1)(l_2m_2)\cdots(l_nm_n)\rangle \quad (6.58)$$

occurs in (6.57) with a coefficient whose m-dependence is contained in the CG product

$$(L\Lambda\,|\,L_{n-1}M_{n-1}, l_nm_n)(L_{n-1}M_{n-1}\,|\,L_{n-2}M_{n-2}, l_{n-1}m_{n-1}) \times\cdots\times(L_2M_2\,|\,l_1m_1, l_2m_2). \quad (6.59)$$

This procedure can be repeated with Λ as well as all m and M quantum numbers reversed. According to the first definition of $| \nu -\Lambda\, SM_S\rangle$, the coefficient of

$$| (l_1 -m_1)(l_2 -m_2)\cdots(l_n -m_n)\rangle$$

would be the same as that of (6.58). However, on reversing all the magnetic quantum numbers in (6.59), we introduce, by Eq. (1.31) and the properties of the 3-j symbols, the phase $(-1)^x$, where

$$\begin{aligned} x &= (L_{n-1} + l_n - L) + (L_{n-2} + l_{n-1} - L_{n-1}) + \cdots + (l_1 + l_2 - L_2) \\ &= \sum_t l_t - L. \end{aligned}$$

This is not, in general, even; and hence the two definitions are not equivalent. The second one leads to

$$R \,|\, \nu L\Lambda SM_S\rangle = (-1)^{\Sigma l - L + \Lambda} \,|\, \nu L -\Lambda\, SM_S\rangle.$$

We shall not, however, adopt this for the simple reason that it is not obvious, except in certain very special cases, what values of l_t and L should be picked for a molecular state. We notice, in passing, that Σl specifies the parity of the atomic configuration $(l_1 l_2 \cdots l_n)$.

The above analysis is nevertheless useful when we consider the state $| \nu\Lambda S\Sigma\rangle$ in Hund's case (a) coupling. For S, unlike L, is a good quantum number, and so $| \nu\Lambda S -\Sigma\rangle$ must be defined by interchanging α and β, and including the overall phase factor $(-1)^y$, where

$$y = \sum_t s_t - S = \tfrac{1}{2}n - S.$$

Under the action of R, the spin functions transform as in Eqs. (6.54). Since $\frac{1}{2}n - \Sigma$ is the number of β spins in each determinantal product state contributing to $| \nu\Lambda S\Sigma\rangle$, we have

$$R \,|\, S\Sigma\rangle = (-1)^{\frac{1}{2}n - \Sigma - \frac{1}{2}n + S} \,|\, S -\Sigma\rangle = (-1)^{S-\Sigma} \,|\, S -\Sigma\rangle.$$

Combining this result with the orbital transformation properties (as given in Eq. (6.56)), we get, for $\Lambda \neq 0$,

$$R \,|\, \nu\Lambda S\Sigma\rangle = (-1)^{S-\Sigma-\Lambda} \,|\, \nu -\Lambda S -\Sigma\rangle. \qquad (6.60)$$

The integral nature of Λ allows us to reverse its sign in the phase factor.

6.10 INTRINSIC PARITY

For molecules for which $Z_A = Z_B$, the Hamiltonian H_e is invariant under inversion in the midpoint of the internuclear axis. The orbitals can thus be classified by their intrinsic parity: that is, whether they are odd or even (u or g) with respect to the inversion operation. The total electronic eigenfunctions $|\nu\Lambda S\Sigma\rangle$ and $|\nu\Lambda SM_S\rangle$ can also be assigned intrinsic parity labels. Following the discussion of Eqs. (6.53), it is immaterial whether the spin parts are subjected to the inversion operation or not.

As an example, consider the O_2 molecule. In the ground configuration, the orbitals $1\sigma_g$, $1\sigma_u$, $2\sigma_g$, $2\sigma_u$, $3\sigma_g$ and $1\pi_u$ are fully occupied [77, App. 11]; that is, the σ orbitals (distinguished by a numerical prefix) contain pairs of electrons with opposing spins, while the π orbital, for which $m = \pm 1$, contains four electrons. The two remaining belong to $1\pi_g$. The various states of the open-shell configuration $(1\pi_g)^2$ can be distinguished by adding m as a subscript and by assigning plus and minus signs for the spin orientations. There are six possible states in all. Using curly brackets to denote normalized determinantal product states, we obtain

$$\{\overset{+}{\pi}_{g1}\overset{+}{\pi}_{g-1}\} \qquad (\Lambda M_S \equiv 0\,1),$$

$$\{\overset{-}{\pi}_{g1}\overset{-}{\pi}_{g-1}\} \qquad (\Lambda M_S \equiv 0\,-1),$$

$$\{\overset{+}{\pi}_{g1}\overset{-}{\pi}_{g1}\} \qquad (\Lambda M_S \equiv 2\,0),$$

$$\{\overset{+}{\pi}_{g-1}\overset{-}{\pi}_{g-1}\} \qquad (\Lambda M_S \equiv -2\,0),$$

$$\left.\begin{matrix}\{\overset{+}{\pi}_{g1}\overset{-}{\pi}_{g-1}\}\\ \{\overset{+}{\pi}_{g-1}\overset{-}{\pi}_{g1}\}\end{matrix}\right. \qquad (\Lambda M_S \equiv 0\,0).$$

In all these states, the closed-shell structure

$$(1\sigma_g)^2(1\sigma_u)^2(2\sigma_g)^2(2\sigma_u)^2(3\sigma_g)^2(1\pi_u)^4 \tag{6.61}$$

is understood. The states for which $(\Lambda M_S) \equiv (\pm 2\,0)$ prove the existence of a ${}^1\Delta$ term; the pair corresponding to $(\Lambda M_S) \equiv (0\,\pm 1)$ show that a ${}^3\Sigma$ term also exists. The prefixed superscripts to Δ and Σ specify the multiplicity $2S + 1$. Since only two of the three states belonging to ${}^3\Sigma$ have been accounted for, the remaining component must be a linear superposition of the two states for which $(\Lambda M_S) \equiv (0\,0)$. The orthogonal combination can only belong to ${}^1\Sigma$. As both π electrons possess even parity, the terms are ${}^1\Delta_g$, ${}^3\Sigma_g$ (the ground term), and ${}^1\Sigma_g$.

Let us investigate whether the ${}^3\Sigma_g$ term is ${}^3\Sigma_g^+$ or ${}^3\Sigma_g^-$. Writing out the determinantal product state in detail, we get, on neglecting the inert configuration (6.61),

$$\{\overset{+}{\pi}_{g1}\overset{+}{\pi}_{g-1}\} = \sqrt{\tfrac{1}{2}}\,[\pi_{g1}(1)\pi_{g-1}(2) - \pi_{g-1}(1)\pi_{g1}(2)]\alpha(1)\alpha(2).$$

From Eq. (6.55),

$$R\pi_{g1} = -\pi_{g-1}, \qquad R\pi_{g-1} = -\pi_{g1}.$$

Thus, on applying R to the orbitals, we obtain

$$\{\overset{+}{\pi}_{g1}\overset{+}{\pi}_{g-1}\} \rightarrow \sqrt{\tfrac{1}{2}}\,[\pi_{g-1}(1)\pi_{g1}(2) - \pi_{g1}(1)\pi_{g-1}(2)]\alpha(1)\alpha(2)$$

$$= \{\overset{+}{\pi}_{g-1}\overset{+}{\pi}_{g1}\} = -\{\overset{+}{\pi}_{g1}\overset{+}{\pi}_{g-1}\},$$

showing that the term in question is, in fact, ${}^3\Sigma_g^-$. The above analysis in no way depends on the axis of quantization for electron spin. If this axis is the internuclear axis (in which case M_S would be replaced by Σ), we can easily apply R to both orbital and spin functions, getting

$$R\{\overset{+}{\pi}_{g1}\overset{+}{\pi}_{g-1}\} = -\{\overset{-}{\pi}_{g1}\overset{-}{\pi}_{g-1}\}.$$

It should be noted that we have nowhere verified that $\{\overset{+}{\pi}_{g1}\overset{+}{\pi}_{g-1}\}$ is an eigenfunction of H_e; the best that we can hope for is that it is a good approximation to it. In the absence of any contributions to H_e from spin interactions, all admixtures into ${}^3\Sigma_g^-$ must themselves possess the symmetries implied by the state labels.

6.11 PARITY

We turn now from the electronic Hamiltonian to the total Hamiltonian H. It is evident from Eq. (6.27) that the total kinetic energy T is invariant under the operations

$$\mathbf{r}_A \rightarrow -\mathbf{r}_A, \qquad \mathbf{r}_B \rightarrow -\mathbf{r}_B, \qquad \mathbf{r}_t \rightarrow -\mathbf{r}_t,$$

which invert the entire molecule in the origin of coordinates. In the absence of external fields, all potential-energy terms depend on the distances between particles: they too must be invariant under the inversion operation I. All additions to H from the effects of electron spin are similarly invariant, as can be seen by looking forward to Sections 7.2 and 7.4, and making the replacements $\boldsymbol{l}_t \rightarrow \boldsymbol{l}_t$ and $\mathbf{s}_t \rightarrow \mathbf{s}_t$ for electron t. This means that the total molecular eigenfunction must be susceptible of a parity classification.

We have now to thread our way through the two coordinate transformations (6.4) and (6.8) to see what inversion implies. The first is easy:

$$\mathbf{r}_O \to -\mathbf{r}_O, \qquad \mathbf{r}_{BA} \to -\mathbf{r}_{BA}, \qquad \mathbf{r}_{tC} \to -\mathbf{r}_{tC}.$$

We begin work on the second transformation by noting that the replacement $\mathbf{r}_{BA} \to -\mathbf{r}_{BA}$ leads to

$$\theta \to \pi - \theta, \qquad \phi \to \pi + \phi.$$

Provided that the role of the added variable is preserved, we can assign it any transformation we wish. It is convenient to take $\gamma \to \pi - \gamma$. In this case, Eqs. (1.10) can be used to give

$$\xi_t \to \xi_t, \qquad \eta_t \to -\eta_t, \qquad \zeta_t \to \zeta_t, \tag{6.62}$$

corresponding to a reflection R in the $\xi\zeta$ plane. At the same time, $-i\,\partial/\partial\gamma$ and (6.19) change sign, so γ continues to serve the purpose for which it was introduced in Section 6.3. As for the rotation eigenfunctions, we can use Eqs. (1.12)–(1.15) and (3.7) to derive

$$|JM_J\,-\Omega\rangle \to (-1)^{J+\Omega}\,|JM_J\Omega\rangle \qquad (\text{case (a)})$$

$$|NM\,-\Lambda\rangle \to (-1)^{N+\Lambda}\,|NM\Lambda\rangle \qquad (\text{case (b)}).$$

For Hund's case (a), the reflection operation R in the molecular frame F' yields

$$|\nu\Lambda S\Sigma\rangle \to (-1)^q\,|\nu\,-\Lambda S\,-\Sigma\rangle,$$

where, from Eq. (6.60), $q = S - \Sigma - \Lambda$ when $\Lambda \neq 0$. For the state Σ^-, q should be augmented by 1; no adjustment is necessary for Σ^+. For Hund's case (b), Eqs. (6.56) show that

$$|\nu\Lambda SM_S\rangle \to (-1)^{q'}\,|\nu\,-\Lambda SM_S\rangle,$$

where $q' = -\Lambda$ if $\Lambda \neq 0$; otherwise it is 0 for a Σ^+ state and 1 for a Σ^- state.

Before assembling the total states, we note that the vibrational variable r is unchanged under inversion, since it is simply $|\mathbf{r}_{AB}|$. Thus

$$I\,|[\nu\Lambda S\Sigma][v][JM_J\,-\Omega]\rangle = (-1)^{J+S+w}\,|[\nu\,-\Lambda\;S\,-\Sigma][v][JM_J\Omega]\rangle,$$

$$I\,|[\nu\Lambda SM_S][v][NM\,-\Lambda]\rangle = (-1)^{N+w}\,|[\nu\,-\Lambda\;SM_S][v][NM\Lambda]\rangle,$$

where $w = 0$ for all electronic states except Σ^-, in which case it is 1. To form states of definite parity, we need the combinations

$$|a\pm\rangle = \sqrt{\tfrac{1}{2}}\,|[\nu\Lambda S\Sigma][v][JM_J\,-\Omega]\rangle \pm \sqrt{\tfrac{1}{2}}\,|[\nu\,-\Lambda\;S\,-\Sigma][v][JM_J\Omega]\rangle, \tag{6.63}$$

$$|b\pm\rangle = \sqrt{\tfrac{1}{2}}\,|[\nu\Lambda SM_S][v][NM\,-\Lambda]\rangle \pm \sqrt{\tfrac{1}{2}}\,|[\nu\,-\Lambda\;SM_S][v][NM\Lambda]\rangle, \tag{6.64}$$

provided, of course, that the two kets of a pair are distinct. Of those special cases where they are not, either one, by itself, possesses definite parity. It is interesting to note that any one of the four linear combinations above alternates in parity with successive angular momenta (J or N).

Further symmetry properties are exhibited by molecules containing two identical nuclei. The Hamiltonian must in this case be invariant with respect to the substitutions $\mathbf{r}_A \to \mathbf{r}_B$ and $\mathbf{r}_B \to \mathbf{r}_A$, which imply $\mathbf{r}_{AB} \to -\mathbf{r}_{AB}$. Just as in the previous case, this replacement leads to

$$\theta \to \pi - \theta, \qquad \phi \to \pi + \phi.$$

However, the vectors $\mathbf{r}_{tC}$ are now unchanged; so the analogs of the substitutions (6.62) turn out to be

$$\xi_t \to -\xi_t, \qquad \eta_t \to \eta_t, \qquad \zeta_t \to -\zeta_t.$$

Evidently the entire analysis of the previous paragraph can be repeated after the inversions $\boldsymbol{\varrho}_t \to -\boldsymbol{\varrho}_t$ have been performed. These purely electronic inversions in the center of the molecular frame can only affect the orbital parts of the electronic eigenfunctions $|\nu\Lambda S\Sigma\rangle$ or $|\nu\Lambda SM_S\rangle$; and it is precisely to describe how such functions behave under inversion that the labels g and u were introduced in Section 6.10. To put this on an algebraic basis, we define the intrinsic parity symbol p to be 0 for a g state and 1 for a u state. The states $|a\pm\rangle$ and $|b\pm\rangle$ of Eqs. (6.63)–(6.64) are eigenfunctions of the operation O that interchanges the two nuclei; in fact, we can quickly verify that

$$O\,|a\pm\rangle = (-1)^{J+S+p+\frac{1}{2}\mp\frac{1}{2}+w}\,|a\pm\rangle \qquad (\Lambda \neq 0 \quad \text{or} \quad S \neq 0), \quad (6.65)$$

$$O\,|b\pm\rangle = (-1)^{N+p+\frac{1}{2}\mp\frac{1}{2}}\,|b\pm\rangle \qquad (\Lambda \neq 0). \quad (6.66)$$

For the special cases in which the two kets that comprise $|a\pm\rangle$ or $|b\pm\rangle$ are identical, we have

$$O\,|[\nu\,{}^1\Sigma_p{}^\pm][v][JM_J0]\rangle = (-1)^{J+\frac{1}{2}\mp\frac{1}{2}+p}\,|[\nu\,{}^1\Sigma_p{}^\pm][v][JM_J0]\rangle, \quad (6.67)$$

$$O\,|[\nu\,{}^{2S+1}\Sigma_p{}^\pm][v][NM0]\rangle = (-1)^{N+\frac{1}{2}\mp\frac{1}{2}+p}\,|[\nu\,{}^{2S+1}\Sigma_p{}^\pm][v][NM0]\rangle. \quad (6.68)$$

The labels s and a (for symmetric and antisymmetric) are sometimes applied to states whose eigenvalues of O are $+1$ and -1 respectively.

So far, a nucleus has been regarded merely as a point particle characterized by a mass and an electric charge. In actual fact, most nuclei possess a nonzero angular momentum $\mathbf{I}$ (the so-called nuclear spin). Even if there are no interaction terms involving $\mathbf{I}$, the value of the corresponding quantum number I has a profound effect on the acceptable electronic states of a homonuclear diatomic molecule. The reason for this stems from the fact that nucleons (both protons and neutrons) possess spins of $\frac{1}{2}$. Since their purely orbital contributions to I must necessarily be integral, the integral or half-integral character of I indicates an even or odd number of nucleons. Being fermions, the total eigenfunction of a system of two nuclei must be antisymmetric with respect to the interchange of two protons or two neutrons. If there are A nucleons in either nucleus, the interchange of the two nuclei should produce a phase factor $(-1)^A$. This number is $+1$ if I is integral and -1 if I is half-integral. The spatial part of the nuclear eigenfunction is already contained in the states we have been investigating; the spin part is conveniently written in the coupled form $|(II)TM_T\rangle$, where T is the combined nuclear spin. Under a nuclear interchange,

$$|(II)TM_T\rangle \to (-1)^{2I-T}|(II)TM_T\rangle, \tag{6.69}$$

the phase factor coming directly from Eq. (1.32). This phase has now to be combined with the phases of Eqs. (6.65)–(6.68) to produce $+1$ (for integral I) or -1 (for half-integral I).

The ground term ${}^3\Sigma_g^-$ of the O_2 molecule provides an interesting example. For O^{16}, the nuclear ground state corresponds to $I = 0$, and so $T = 0$. From Eq. (6.68), we have

$$(-1)^{N+\frac{1}{2}+\frac{1}{2}+0} = +1,$$

so for ${}^3\Sigma_g^-$ the only rotational levels to occur are those for which N is odd.

PROBLEMS

6.1 Prove that, for rotating the spatial coordinates of the electrons of a diatomic molecule, the operator $D(\omega)$ of Section 1.3 can be written as

$$D(\omega)^{-1} = \exp i\gamma L_z \exp i\theta L_y \exp i\phi L_z = \exp i\gamma L_\zeta \exp i\theta L_\eta \exp i\phi L_\zeta,$$

in the notation of Section 6.2. Show that Eqs. (6.13) can be derived from

$$\partial/\partial\theta_{BA} = D(\omega)^{-1}(\partial/\partial\theta)D(\omega),$$

$$\partial/\partial\phi_{BA} = D(\omega)^{-1}(\partial/\partial\phi)D(\omega).$$

(Transformations of this type have been given by Kronig [99]; his S is our $D(\omega)$.)

6.2 A fictitious nucleus of mass M_0 is added to a diatomic molecule. Show that, in the limit $M_0 \to 0$, the Hamiltonian contains a term proportional to

$$(N_\zeta - L_\zeta)^2/I_3$$

in which $I_3 = M_0\rho^2$, where ρ is the distance of the fictitious nucleus from the nuclear axis AB. (This divergent term contributes nothing to the energy provided Eq. 6.22 holds.)

6.3 Prove that the operators

$$l_x = \frac{1}{2}\left(r\frac{d^2}{dr^2} - r - \frac{c}{r}\right), \qquad l_y = -r\frac{d}{dr}, \qquad l_z = \frac{1}{2}i\left(r\frac{d^2}{dr^2} + r - \frac{c}{r}\right)$$

satisfy the usual commutation relations for an angular momentum vector. Show also that $\boldsymbol{l}^2 = c$.

The Hamiltonian H for a rotating oscillator is given by

$$H = -\frac{d^2}{dr^2} + \frac{c}{r^2} - \frac{2b}{r}.$$

Prove that $r(H - E)$, where E is the eigenvalue of H, can be expressed as a linear combination of the components of $\boldsymbol{l}$. Find the angle of tilt β for which

$$\exp[-i\beta l_y](rH - rE)\exp[i\beta l_y]$$

depends only on l_z. Hence show that, if n is the (possibly nonintegral) eigenvalue of l_z, then $E = -b^2/n^2$. Obtain the acceptable sequence of l_z values by solving $l(l+1) = c$, and hence obtain Eq. (6.45) for the energy levels of a rotating oscillator. (This general algebraic method is due to Barut [100].)

6.4 If Eq. (6.46) is written in the general form

$$\epsilon_0/b^2 = A + \omega(v + \tfrac{1}{2}) - \omega x(v + \tfrac{1}{2})^2 + BN(N+1) - CN^2(N+1)^2,$$

it is easy to confirm that

$$C\omega^2 = 4B^3$$

for the potential $W(r)$ of Eq. (6.42). Show that, for small oscillations, the same result holds for the Morse potential.

6.5 The time-reversal operator T has the effect of changing the sign of any angular momentum vector $\mathbf{J}$. Show that the standard relations (1.3) for the associated kets $|JM\rangle$ are maintained if we take $TiT^{-1} = -i$ and choose phases so that

$$T|JM\rangle = (-1)^{J-M}|J\,-M\rangle.$$

Prove that the rotational kets of Eq. (3.7) transform according to

$$T|JM\,-N\rangle = (-1)^{M+N}|J\,-MN\rangle.$$

Deduce that, for the states of Eqs. (6.63)–(6.64),

$$T^2 \mid a\pm\rangle = (-1)^{2S} \mid a\pm\rangle,$$
$$T^2 \mid b\pm\rangle = (-1)^{2S} \mid b\pm\rangle$$

(see Heine [101]).

6.6 The generators for $R(4)$ are given as $\mathbf{N}^{(10)}$ and $\mathbf{G}^{(01)}$ in Section 6.4 for a spin-free diatomic molecule. The total spin $\mathbf{K}$ of a certain subset of particles is referred to the molecular frame F', thereby ensuring that K_ξ, K_η, and K_ζ commute with $\mathbf{N}$. Show that the components K_x, K_y, and K_z behave as a vector with respect to $\mathbf{N}^{(10)}$ but commute with $\mathbf{G}^{(01)}$. Deduce that the adjustments

$$\mathbf{N}^{(10)} \rightarrow \mathbf{N}^{(10)}, \qquad \mathbf{G}^{(01)} \rightarrow \mathbf{G}^{(01)} + \mathbf{K}^{(01)}$$

are necessary to the generators of $R(4)$ if the ranks of $\mathbf{K}^{(10)}$ and $\mathbf{K}^{(01)}$ are to correctly describe the transformation properties of the vectors. Check that these new generators become $\mathbf{J}^{(10)}$ and $\mathbf{P}^{(01)} - \mathbf{J}^{(01)}$ when $\mathbf{K} \equiv \mathbf{S}$, in the notation of Section 6.8.

If, on the other hand, it is supposed that K_x, K_y, and K_z commute with $\mathbf{N}$ (in which case K is referred to F), show that the adjustments to the generators are given by

$$\mathbf{N}^{(10)} \rightarrow \mathbf{N}^{(10)} + \mathbf{K}^{(10)}, \qquad \mathbf{G}^{(01)} \rightarrow \mathbf{G}^{(01)}.$$

6.7 A diatomic molecule is hypothetically formed from an atom by dividing the atomic nucleus into two constituent nuclei and then separating them. Show that an atomic term, characterized by S, L, and a parity p, separates into molecular states whose Λ values are given by

$$|\Lambda| = 0, 1, 2, \ldots, L,$$

and whose total spin quantum numbers are all S. Show that the Σ state in the above sequence is Σ^+ or Σ^- according as $L + p$ is an even or an odd integer. If the two constituent nuclei are identical, show that the molecular states can be classified as g or u according as $p = 0$ or 1.

Suppose, on the other hand, that the molecule is formed by bringing together two atoms whose terms are characterized by $(S_A L_A p_A)$ and $(S_B L_B p_B)$. Prove that the molecular states $S \mid \Lambda \mid$ provided by these two terms can be found by forming all possible couplings $(S_A S_B)S$ and $(L_A L_B) L$, and then letting Λ run over all M_L values. Show that the number $N(\Sigma)$ of Σ states is the lesser of $2L_A + 1$ and $2L_B + 1$, and that

$$N(\Sigma^+) - N(\Sigma^-) = \pm 1,$$

where the upper sign is to be taken if $p_A + p_B + L_A + L_B$ is even, and the lower sign if it is odd. (Questions of this kind were first discussed by Wigner and Witmer [102]).

7

The Hydrogen Molecule

7.1 INTRODUCTION

The treatment of diatomic molecules has so far neglected terms in the Hamiltonian that involve electron spin. Their inclusion considerably increases the complexity of the analysis. It seems best for our purposes to make continual reference to a concrete example. In this way the general method can be appreciated without our having to specify the formalism in all its detail. The lowest $^3\Pi_u$ state of the molecule H_2 serves our purposes well. It is, in fact, the lowest bound triplet state, though a $^3\Sigma_g{}^+$ lies close [46]. Several of the rotational components of $^3\Pi_u$ have been shown to be metastable, and this permits their splittings to be determined by resonance methods [103–105]. Optical spectra have also been analyzed [106, 107], and the experimental data have stimulated considerable theoretical work [108–112]. The analysis that follows is essentially a reworking of that of Fontana [108] and Chiu [110].

Before writing down the new contributions to the Hamiltonian, the existence of two forms of H_2 should be mentioned. Each nucleus of H_2 comprises a single proton with spin $\frac{1}{2}$. When the total spin T is 1, the molecule is spoken of as ortho-hydrogen; when $T = 0$, it is referred to as para-hydrogen. From Eqs. (6.66) and (6.69), we must have

$$(-1)^{N+1+\frac{1}{2}\mp\frac{1}{2}}(-1)^{1-T} = -1,$$

assuming (as is the case) that Hund's case (b) is the best representation of the intramolecular coupling. Thus, for para-H_2 we take the states $|\,b+\rangle$ when N is odd, and the states $|\,b-\rangle$ when N is even. The opposite choice must be made for ortho-H_2. In view of the fact that no mechanism has been postulated for coupling the two nuclear spins, it may seem strange that their resultant T should play such a crucial role in determining the acceptable states $|\,b+\rangle$ or $|\,b-\rangle$. The reason is that the symmetry or antisymmetry of the spin part of the nuclear eigenfunction is exhibited directly through T. In a similar way, the total spin S of a pair of equivalent electrons in an atom determines the acceptable values of L even in the absence of any spin terms in the Hamiltonian.

7.2 THE EFFECTIVE FINE-STRUCTURE HAMILTONIAN

The first-order effects of terms involving the interactions between particles of spin $\frac{1}{2}$ are obtained by taking the nonrelativistic limit of the Breit interaction [113]. The terms break up into those independent of spin; those linear in the spins of the particles; and those quadratic in the spins. Our present interest centers on the terms linear in the spins: these include what are usually referred to as spin–orbit and spin–other-orbit interactions. For the moment, all terms involving nuclear spins are dropped, since they properly come under the heading of hyperfine interactions. We are left with the following contributions to the Hamiltonian:

$$H_{\mathrm{so}} = 2\beta^2 \sum_{t\neq v} \left[\nabla_t \left(\frac{1}{r_{tv}}\right) \times \mathbf{p}_t\right]\cdot(\mathbf{s}_t + 2\mathbf{s}_v) - 2\beta^2 \sum_{tX} \left[\nabla_t \left(\frac{Z_X}{r_{tX}}\right) \times \mathbf{p}_t\right]\cdot\mathbf{s}_t, \tag{7.1}$$

$$H_{\mathrm{sr}} = -4\beta\beta_N \sum_{tX} \left[\nabla_X \left(\frac{Z_X}{r_{Xt}}\right) \times \mathbf{p}_X\right]\cdot\mathbf{s}_t. \tag{7.2}$$

In these expressions, the nuclear index X runs over the two possibilities A and B, and the electronic indices t and v can be 1 and 2. The Bohr magneton β is $e\hbar/2mc$; the nuclear magneton β_N is $e\hbar/2M_Pc$, where M_P

is the mass of the proton. It is perhaps worth noting that

$$4\beta^2 = \frac{e^2\hbar^2}{m^2c^2} = \left(\frac{e^2}{c\hbar}\right)^2\left(\frac{\hbar^2}{me^2}\right)^3\left(\frac{e^4m}{\hbar^2}\right) = \alpha^2 a_0{}^3\epsilon_0,$$

where α is the fine-structure constant ($\simeq 1/137$) and ϵ_0 is one atomic unit of energy (equivalent to 219475 cm^{-1}).

To begin with, we immediately set aside H_{sr}. This term involves the nuclear momenta $\mathbf{p}_X$, and gives rise to the coupling between spin and rotation (see Section 7.6). To simplify H_{so}, we use the fact that our intended application is to the $^3\Pi$ term of H_2. For this,

$$\langle {}^3\Pi \mid (\mathbf{s}_1 - \mathbf{s}_2) \mid {}^3\Pi \rangle = 0,$$

since $^3\Pi$ must be symmetric with respect to the interchange $1 \leftrightarrow 2$ in spin space, whereas $\mathbf{s}_1 - \mathbf{s}_2$ is antisymmetric. We may therefore write

$$\mathbf{s}_t + 2\mathbf{s}_v = \tfrac{3}{2}(\mathbf{s}_t + \mathbf{s}_v) - \tfrac{1}{2}(\mathbf{s}_t - \mathbf{s}_v) \equiv \tfrac{3}{2}\mathbf{S},$$
$$\mathbf{s}_t = \tfrac{1}{2}(\mathbf{s}_t + \mathbf{s}_v) + \tfrac{1}{2}(\mathbf{s}_t - \mathbf{s}_v) \equiv \tfrac{1}{2}\mathbf{S}.$$

We get, in all,

$$H_{\text{so}} = \beta^2\left[-\frac{3}{r_{12}{}^3}(\mathbf{r}_{12} \times \mathbf{p}_1) + \frac{3}{r_{12}{}^3}(\mathbf{r}_{12} \times \mathbf{p}_2) + \frac{Z_A}{r_{1A}{}^3}(\mathbf{r}_{1A} \times \mathbf{p}_1)\right.$$
$$\left. + \frac{Z_A}{r_{2A}{}^3}(\mathbf{r}_{2A} \times \mathbf{p}_2) + \frac{Z_B}{r_{1B}{}^3}(\mathbf{r}_{1B} \times \mathbf{p}_1) + \frac{Z_B}{r_{2B}{}^3}(\mathbf{r}_{2B} \times \mathbf{p}_2)\right]\cdot\mathbf{S}.$$

We now break every position vector up into two parts by writing

$$\mathbf{r}_{12} = \mathbf{r}_{1C} - \mathbf{r}_{2C},$$
$$\mathbf{r}_{1A} = \mathbf{r}_{1C} - \mathbf{r}_{AC} = \mathbf{r}_{1C} - \{M_B/(M_A + M_B)\}\mathbf{r}_{AB},$$

etc. With a little analysis it is not difficult to see that all terms involving $\mathbf{r}_{AB}$ contribute nothing to the final matrix element. Consider, for example, $r_{1A}{}^{-3}(\mathbf{r}_{AB} \times \mathbf{p}_1)$. In the frame F' of the molecule, only the ζ component of $\mathbf{r}_{AB}$ exists; and this means that only $(\mathbf{p}_1)_\xi$ and $(\mathbf{p}_1)_\eta$ appear. From the last of the set of equations (6.5), we see that we can write $\mathbf{p}_t = \mathbf{p}_{tC}$ if purely translational motions of the molecule are ignored. The ξ and η components of $\mathbf{p}_{tC}$ are simply $-i\,\partial/\partial\xi_t$ and $-i\,\partial/\partial\eta_t$. Being shift operators, they obey the selection rule $\Delta\Lambda = \pm 1$. (The proof of this follows the derivation of Eq. (6.32).) The term $r_{1A}{}^{-3}$ is scalar with respect to L_ζ, so this selection rule holds for $r_{1A}{}^{-3}(\mathbf{r}_{AB} \times \mathbf{p}_1)$ itself. Since a Π state is a linear combination of the two states for which $\Lambda = \pm 1$, we see that, for a stationary center of mass O,

$$\langle {}^3\Pi_u \mid r_{tX}^{-3}(\mathbf{r}_{AB} \times \mathbf{p}_t) \mid {}^3\Pi_u \rangle = 0.$$

The expression for H_{so} now simplifies to

$$H_{so} = \beta^2 \mathbf{U} \cdot \mathbf{S}, \tag{7.3}$$

where

$$\mathbf{U} = \left(\frac{Z_A}{r_{1A}^3} + \frac{Z_B}{r_{1B}^3} - \frac{3}{r_{12}^3}\right) \boldsymbol{l}_{1C} + \left(\frac{Z_A}{r_{2A}^3} + \frac{Z_B}{r_{2B}^3} - \frac{3}{r_{12}^3}\right) \boldsymbol{l}_{2C}$$

$$+ 3\left(\frac{1}{r_{12}^3}\right)(\mathbf{r}_{1C} \times \mathbf{p}_2 + \mathbf{r}_{2C} \times \mathbf{p}_1). \tag{7.4}$$

We have written $\boldsymbol{l}_{tC}$ for $\mathbf{r}_{tC} \times \mathbf{p}_t$ here. The expression (7.3) for H_{so} constitutes an effective Hamiltonian which we may use within states described by ${}^3\Pi_u$. The cross terms of the type $\mathbf{r}_{tC} \times \mathbf{p}_v$ (with $t \neq v$) appear to have been overlooked in the early theoretical treatments [108]. Some justification for doing so could have been made by introducing the formal assumption that all exchange integrals involving the two electrons of ${}^3\Pi_u$ be dropped. This is not unreasonable for electrons whose orbits only slightly overlap.

7.3 FINE STRUCTURE

To calculate the effects of H_{so}, we assume that the states are given, to a good approximation, by the kets $\mid (NS)J\rangle$, as described in Section 6.8. This corresponds to Hund's case (b); the justification for using this coupling will appear in Section 7.4, when theory is compared to experiment. We can begin the analysis by applying Eq. (1.67):

$$\begin{aligned}\langle (NS)JM \mid H_{so} \mid (NS)J'M' \rangle &= \beta^2 \langle (NS)JM \mid \mathbf{U} \cdot \mathbf{S} \mid (NS)J'M' \rangle \\ &= \beta^2 (-1)^{J+S+N} \delta(J, J') \delta(M, M') \\ &\quad \times (N \| U \| N)(S \| S \| S) \begin{Bmatrix} N & 1 & N \\ S & J & S \end{Bmatrix}.\end{aligned}$$

If we replace the 6-j symbol by its explicit form (see Appendix 1) and use Eq. (1.53) to evaluate the reduced matrix element of $\mathbf{S}$, we get, for the nonvanishing term,

$$\langle (NS)J \mid H_{\text{so}} \mid (NS)J \rangle$$
$$= \beta^2(N \,\|\, U \,\|\, N) \frac{J(J+1) - N(N+1) - S(S+1)}{\{2N(2N+1)(2N+2)\}^{1/2}}. \qquad (7.5)$$

This specifies unambiguously the dependence on J. Since the operator $\mathbf{N}\cdot\mathbf{S}$ would yield precisely the same 6-j symbol, we may write

$$H_{\text{so}} \equiv \lambda \mathbf{N}\cdot\mathbf{S},$$

where λ is independent of J. Like $\beta^2\mathbf{U}\cdot\mathbf{S}$, the expression $\lambda\mathbf{N}\cdot\mathbf{S}$ is an effective Hamiltonian for H_{so}. However, its usefulness is even more limited than that of $\beta^2\mathbf{U}\cdot\mathbf{S}$, since $\lambda\mathbf{N}\cdot\mathbf{S}$ can only be used within a single multiplet for which N and S are constant. Under these conditions, it is easy to show that the *Landé interval rule* is valid: that is, the energy between the pair of levels characterized by J and $J-1$ is proportional to J.

If we want to relate the level spacings for different N, we must turn our attention to the reduced matrix element of $\mathbf{U}$ in Eq. (7.5). There are, in fact, four possible matrix elements here, as we can see by writing out the reduced matrix element in full:

$$(N \,\|\, U \,\|\, N) = ([\nu\Lambda][v][N\,-\Lambda] \,|\!:\, U_{\cdot 0}^{(10)} :\!|\, [\nu\Lambda'][v][N\,-\Lambda']). \qquad (7.6)$$

We have used the notation of Section 3.3 for a matrix element reduced in the laboratory frame F. The assignment of ranks to $\mathbf{U}$ follows from the fact that the spin has already been treated in the laboratory frame F, and so we must interpret $\mathbf{U}\cdot\mathbf{S}$ as $\mathbf{U}^{(10)}\cdot\mathbf{S}^{(10)}$ rather than $\mathbf{U}^{(01)}\cdot\mathbf{S}^{(01)}$. The four possible matrix elements correspond to $\Lambda = \pm 1$, $\Lambda' = \pm 1$. However, the vector $\mathbf{U}$ (whether regarded as $\mathbf{U}^{(01)}$ or $\mathbf{U}^{(10)}$) cannot connect different Λ values, since a step of 2 units is required. This means that we can limit attention to the diagonal cases for which $\Lambda = \Lambda'$. The absence of cross terms implies that there can be no distinction between $|\, b+ \rangle$ and $|\, b- \rangle$, and hence the results of this section are equally valid for para-H_2 and ortho-H_2.

We now use

$$\mathbf{U}^{(10)} = (\mathbf{D}^{(11)}\mathbf{U}^{(01)})^{(10)}. \qquad (7.7)$$

The tensor $\mathbf{D}^{(11)}$ acts solely on the rotational part $|\, [N\,-\Lambda] \rangle$ of the states making up the reduced matrix element of Eq. (7.6). On the other hand, $\mathbf{U}^{(01)}$ involves the relative distances between particles as well as such vectors as $\boldsymbol{l}_{tC}^{(01)}$ and $(\mathbf{r}_{tC}\mathbf{p}_v)^{(01)}$. As such, it is independent of the direction of $\mathbf{r}_{AB}$ and thus of the Euler angles too. It therefore acts solely on the *vibronic* part $|\,[\nu\Lambda][v]\rangle$. To connect identical Λ values, the component $U_{00}{}^{(01)}$ is required. We uncouple in the molecular frame F', getting

$$
\begin{aligned}
(N\,\|\,U\,\|\,N)&\\
&= (10, 10\,|\,00)\,([\nu\Lambda][v][N-\Lambda]\,|\!:D_{\cdot 0}{}^{(11)}U_{00}{}^{(01)}:\!|\,[\nu\Lambda][v][N-\Lambda])\\
&= -(3)^{-1/2}(N-\Lambda\,|\!:D_{\cdot 0}{}^{(11)}:\!|\,N-\Lambda)\,\langle[\nu\Lambda][v]\,|\,U_{00}{}^{(01)}\,|\,[\nu\Lambda][v]\rangle\\
&= -(3)^{-1/2}(-1)^{N+\Lambda}\begin{pmatrix} N & 1 & N\\ \Lambda & 0 & -\Lambda\end{pmatrix}(N\,\|\,D^{(11)}\,\|\,N)\\
&\quad\times\langle[\nu\Lambda][v]\,|\,U_{00}{}^{(01)}\,|\,[\nu\Lambda][v]\rangle.
\end{aligned}
$$

An explicit form for the 3-*j* symbol is given in Appendix 1, and the reduced matrix element of $\mathbf{D}^{(11)}$ can be written down from Eq. (3.12). All in all, we find

$$
\begin{aligned}
\langle (NS)J\,|\,H_{\mathrm{so}}\,|\,(NS)J\rangle&\\
&= \frac{\Lambda\beta^2}{2N(N+1)}\{J(J+1) - N(N+1) - S(S+1)\}\\
&\quad\times\langle[\nu\Lambda][v]\,|\,U_{00}{}^{(01)}\,|\,[\nu\Lambda][v]\rangle. \qquad (7.8)
\end{aligned}
$$

The N dependence, as well as the J dependence, is exhibited in this formula. As both Λ and the matrix element of $U_{00}{}^{(01)}$ change sign under the substitution $\Lambda \rightarrow -\Lambda$, the twofold degeneracy associated with the reflection operation R is maintained. To the order at which we are working, each level possesses a degeneracy of $2(2J+1)$.

7.4 SPIN–SPIN INTERACTION

Even in the absence of any information about the matrix element of $U_{00}{}^{(01)}$, we can test Eq. (7.8) by seeing how the dependence of the fine-structure splittings on N and J compares with experiment [103–107]. Much of the fine structure in the rotational levels of the $^3\Pi_u$ state has yet to be resolved, though Lichten and his colleagues have found the splittings of the $N = 2$ level of para-H_2 and those of the $N = 1$ level of ortho-H_2 by resonance methods [104, 105]. Such data as there are make it obvious that there is severe disagreement between theory and experiment. Perhaps the most obvious discrepancy lies in the fact that the observed splittings for the ortho and para forms are not identical; nor do they correspond to the calculations in any reasonably direct way. For example, not even the order of the three levels making up the $N = 2$ multiplet is correctly predicted.

The situation closely parallels the 1s2p ^{3}P multiplet of the helium atom, where the Landé interval rule is badly perturbed (see Problem 7.3). As in the atomic case, the explanation for the discrepancies lies in the neglect of the spin-spin interaction, H_{ss}. This constitutes the part of the Breit interaction that is quadratic in the electron spins. In the nonrelativistic limit,

$$H_{ss} = 4\beta^2 \left[\frac{\mathbf{s}_1 \cdot \mathbf{s}_2}{r_{12}^3} - \frac{3(\mathbf{r}_{12} \cdot \mathbf{s}_1)(\mathbf{r}_{12} \cdot \mathbf{s}_2)}{r_{12}^5} \right]. \tag{7.9}$$

This is identical to the classical interaction between two magnetic dipoles $-2\beta\mathbf{s}_1$ and $-2\beta\mathbf{s}_2$. The contact term involving $\delta(\mathbf{r}_{12})$ does not contribute to the splittings of the multiplets, and it has therefore been dropped. We have now to cast H_{ss} in tensorial form and carry out a similar analysis to that performed for H_{so}.

Since we shall again be working with states corresponding to Hund's case (b), we need to recouple H_{ss} in order to match the description $|(NS)J\rangle$. That is, we need to separate the part acting in the space of $\mathbf{N}$ from that acting in the space of $\mathbf{S}$. Only the second term in H_{ss} presents any difficulty. We write

$$\begin{aligned}(\mathbf{r}_{12} \cdot \mathbf{s}_1)(\mathbf{r}_{12} \cdot \mathbf{s}_2) &= 3(\mathbf{r}_{12}\mathbf{s}_1)^{(0)}(\mathbf{r}_{12}\mathbf{s}_2)^{(0)} \\ &= 3 \sum_{kt} ((11)0, (11)0, 0 \mid (11)k, (11)t, 0) \\ &\quad \times \{(\mathbf{r}_{12}\mathbf{r}_{12})^{(k)}(\mathbf{s}_1\mathbf{s}_2)^{(t)}\}^{(0)}.\end{aligned}$$

The recoupling coefficient is related to a 9-j symbol (see Section 1.11). Since it possesses a column of zeros, it is particularly simple to evaluate. We find

$$((11)0, (11)0, 0 \mid (11)k, (11)t, 0) = \tfrac{1}{3}\delta(k, t)(2k+1)^{1/2}.$$

So

$$(\mathbf{r}_{12} \cdot \mathbf{s}_1)(\mathbf{r}_{12} \cdot \mathbf{s}_2) = \sum_k (2k+1)^{1/2}\{(\mathbf{r}_{12}\mathbf{r}_{12})^{(k)}(\mathbf{s}_1\mathbf{s}_2)^{(k)}\}^{(0)}.$$

Only three values of k are allowed. For $k = 0$,

$$\begin{aligned}(\mathbf{r}_{12}\mathbf{r}_{12})^{(0)} &= -(3)^{-1/2}(\mathbf{r}_{12} \cdot \mathbf{r}_{12}) = -(3)^{-1/2}r_{12}^2, \\ (\mathbf{s}_1\mathbf{s}_2)^{(0)} &= -(3)^{-1/2}(\mathbf{s}_1 \cdot \mathbf{s}_2).\end{aligned}$$

Thus the $k = 0$ term gives $\frac{1}{3}r_{12}^2(\mathbf{s}_1 \cdot \mathbf{s}_2)$, which, when multiplied by 3, exactly cancels the first term in H_{ss}. As for $k = 1$, we know $(\mathbf{r}_{12}\mathbf{r}_{12})^{(1)} = 0$, since this is merely the statement that the vector product of two identical vectors (whose components commute with one another) is null. We are left with $k = 2$. Thus we get, in all,

$$H_{ss} = -12\beta^2(5)^{1/2}r_{12}^{-5}\{(\mathbf{r}_{12}\mathbf{r}_{12})^{(2)}(\mathbf{s}_1\mathbf{s}_2)^{(2)}\}^{(0)}. \tag{7.10}$$

An alternative form can be obtained by writing

$$\mathbf{r}_{12} = r_{12}\mathbf{C}^{(1)}(\theta_{12}\phi_{12})$$

and then using Eq. (2.30) to give

$$(\mathbf{C}^{(1)}\mathbf{C}^{(1)})^{(2)} = (2/3)^{1/2}\mathbf{C}^{(2)}.$$

We now have

$$H_{ss} = -4(6)^{1/2}\beta^2 r_{12}^{-3}(\mathbf{C}^{(2)}(\theta_{12}\phi_{12})\cdot(\mathbf{s}_1\mathbf{s}_2)^{(2)}). \tag{7.11}$$

A similar analysis to that of Section 7.3 can be immediately carried out. The dependence on J of the matrix elements of H_{ss} is given by

$$\langle(NS)J \mid H_{ss} \mid (NS)J\rangle = -4(6)^{1/2}\beta^2(-1)^{N+S+J}\begin{Bmatrix} N & 2 & N \\ S & J & S \end{Bmatrix}$$
$$\times (N \,\|\, r_{12}^{-3}C^{(2)}(\theta_{12}\phi_{12}) \,\|\, N)(S \,\|\, (\mathbf{s}_1\mathbf{s}_2)^{(2)} \,\|\, S). \tag{7.12}$$

The reduced matrix element of $(\mathbf{s}_1\mathbf{s}_2)^{(2)}$ is nonzero for a two-electron system only when $S = 1$. In that case, Eq. (1.62) yields

$$(S \,\|\, (\mathbf{s}_1\mathbf{s}_2)^{(2)} \,\|\, S) = 3(5)^{1/2}\begin{Bmatrix} \frac{1}{2} & \frac{1}{2} & 1 \\ \frac{1}{2} & \frac{1}{2} & 1 \\ 1 & 1 & 2 \end{Bmatrix}(s \,\|\, s \,\|\, s)^2$$
$$= \tfrac{1}{2}(5)^{1/2}.$$

The other reduced matrix element in Eq. (7.12) is completely defined by writing it as

$$([\nu\Lambda][v][N-\Lambda] \,|\!:\, r_{12}^{-3}C_{\cdot 0}^{(20)}(\theta_{12}\phi_{12}) \,:\!|\, [\nu\Lambda'][v][N-\Lambda']). \tag{7.13}$$

The assignment of ranks to $\mathbf{C}$ follows from the fact that the angles $(\theta_{12}\phi_{12})$ are defined in the laboratory frame F, and it is in this frame that the spin part of the matrix element has been reduced. The distance r_{12} is, of course, a scalar in both F and F'. The analogy with Section 7.3 is maintained by taking

$$\mathbf{C}^{(20)}(\theta_{12}\phi_{12}) = \{\mathbf{D}^{(22)}(\phi\theta\gamma)\mathbf{C}^{(02)}(\vartheta_{12}\varphi_{12})\}^{(20)}, \tag{7.14}$$

where the polar angles (ϑ, φ) refer to the frame F'. The matrix element (7.13) becomes

$$(2q, 2\, -q \mid 00)([\nu\Lambda][v][N-\Lambda] \,|\!:\, r_{12}^{-3}D_{\cdot q}^{(22)}C_{0-q}^{(02)} \,:\!|\, [\nu\Lambda'][v][N-\Lambda']), \tag{7.15}$$

where $q = \Lambda' - \Lambda$. The possibility now arises of having $q = \pm 2$; this allows cross terms between the different parts of $| b\pm \rangle$, and hence opens the way for a distinction to be drawn between the ortho and para forms of H_2. Continuing, (7.15) becomes

$$\begin{pmatrix} 2 & 2 & 0 \\ q & -q & 0 \end{pmatrix} (-1)^{N+\Lambda} \begin{pmatrix} N & 2 & N \\ \Lambda & q & -\Lambda' \end{pmatrix} (2N+1)(5)^{1/2}$$

$$\times \langle [\nu\Lambda][v] \,|\, r_{12}^{-3} C_{0-q}^{(02)}(\vartheta_{12}\varphi_{12}) \,|\, [\nu\Lambda'][v] \rangle.$$

All in all,

$$\langle [\nu\Lambda][v][N-\Lambda]SJ \,|\, H_{ss} \,|\, [\nu\Lambda'][v][N-\Lambda']SJ \rangle$$

$$= 2(30)^{1/2}\beta^2(-1)^{J-\Lambda'}(2N+1) \begin{pmatrix} N & 2 & N \\ \Lambda & \Lambda'-\Lambda & -\Lambda' \end{pmatrix} \begin{Bmatrix} N & 2 & N \\ 1 & J & 1 \end{Bmatrix}$$

$$\times \langle [\nu\Lambda][v] \,|\, r_{12}^{-3} C_{0\Lambda-\Lambda'}^{(02)}(\vartheta_{12}\varphi_{12}) \,|\, [\nu\Lambda'][v] \rangle \qquad (7.16)$$

for $S = 1$. This equation specifies the dependence on N and J of the matrix element of H_{ss}.

There remain two distinct matrix elements which we have not yet determined. We write them as

$$B_0 = 4\beta^2 \langle [\nu \pm 1][v] \,|\, r_{12}^{-3} C_{00}^{(02)}(\vartheta_{12}\varphi_{12}) \,|\, [\nu \pm 1][v] \rangle, \qquad (7.17)$$

$$B_2 = 4\beta^2 \langle [\nu \pm 1][v] \,|\, r_{12}^{-3} C_{0\pm 2}^{(02)}(\vartheta_{12}\varphi_{12}) \,|\, [\nu \mp 1][v] \rangle. \qquad (7.18)$$

Thus with the reduced matrix element of **U**, which is conveniently combined with Λ and written as

$$A = \Lambda\beta^2 \langle [\nu \pm 1][v] \,|\, U_{00}^{(01)} \,|\, [\nu \pm 1][v] \rangle, \qquad (7.19)$$

we have in all three unknown quantities in terms of which the fine structure of any rotational level of $^3\Pi_u$ (for a given vibrational quantum number v) can be expressed. If we use explicit expressions for the 3-j and 6-j symbols in Eq. (7.16), we find, using Eq. (6.64), that the energy E_{NSJ} of the level J relative to the center of gravity of the NS multiplet can be concisely expressed as follows:

$$E_{N1\,N+1} = \frac{A}{N+1} - B_0 \frac{3 - N(N+1)}{2(N+1)(2N+3)} \mp B_2 \frac{N}{2N+3} \sqrt{\frac{3}{8}},$$

$$E_{N1N} = -\frac{A}{N(N+1)} + B_0 \frac{3 - N(N+1)}{2N(N+1)} \pm B_2 \sqrt{\frac{3}{8}},$$

$$E_{N1\,N-1} = -\frac{A}{N} - B_0 \frac{3 - N(N+1)}{2N(2N-1)} \mp B_2 \frac{N+1}{2N-1} \sqrt{\frac{3}{8}}, \qquad (7.20)$$

where the upper sign refers to parahydrogen with odd N or orthohydrogen with even N; and the lower sign refers to parahydrogen with even N or orthohydrogen with odd N. If we wish, we can regard these equations as expressing the fine structures of the rotational levels of $^3\Pi_u$ as functions of the three parameters A, B_0, and B_2. In fact, B_0 and B_2 occur through the same linear combination for a given multiplet, and this makes it possible to find A when all three components of a multiplet are known. The data [104, 105] on the $N = 1$ and $N = 2$ levels in ortho-H_2 and para-H_2 respectively lead to $A = -0.124$ cm^{-1} and $A = -0.127$ cm^{-1}. Brooks *et al.* [105] have noted that any difference may correspond to the possibility that the two sets of experimental data may come from two different vibrational states, presumably $v = 0$ and $v = 1$. An analysis by Jette [113a] supports an unpublished revision of the experimental data by Lichten and Vierima, who give $A = -0.121$ cm^{-1} for the $N = 2$ level. The difference from the A value of the $N = 1$ level can then be interpreted as being due to spin-rotation interaction (see Section 7.6 and Problem 7.6). The relevance of this interaction for the excited $^3\Pi_u$ states of hydrogen has been stressed by Miller and Freund [113b].

7.5 THE UNITED-ATOM APPROXIMATION

The quantities A, B_0, and B_2, defined through Eqs. (7.17)–(7.19), can be evaluated only if the electronic structure of the $^3\Pi_u$ state is known. It is often inconvenient—not to say extremely difficult—to make calculations of this kind. A simple device for estimating such integrals as A, B_0, and B_2 is to suppose that the internuclear distance r is zero. The two nuclei coalesce to produce an atom for which, one hopes, the calculations are easier to perform. The actual limiting procedure is of some interest in its own right: the rotational and vibrational spectra expand and go over ultimately to the rotational and vibrational spectra of the atomic nucleus. The purely electronic features are now very much smaller on the energy scale than those coming from vibrations and rotations—the exact opposite to the situation for molecules. The electronic energy levels tend to those of an atom whose nucleus possesses a charge Ze, where $Z = Z_A + Z_B$.

For H_2, the united atom is He. The lowest electronic state possessing the symmetry properties of $^3\Pi_u$ is a component of the term 1s2p ^{3}P. This is a particularly simple term to treat, since the 1s electron is moving almost entirely in the field of an effective nuclear charge $2e$, while the 2p electron, being almost completely screened by the 1s electron, moves in a much more extended orbit corresponding to an effective central charge of e. We can thus use hydrogenic eigenfunctions corresponding to $Z = 2$ (for the 1s

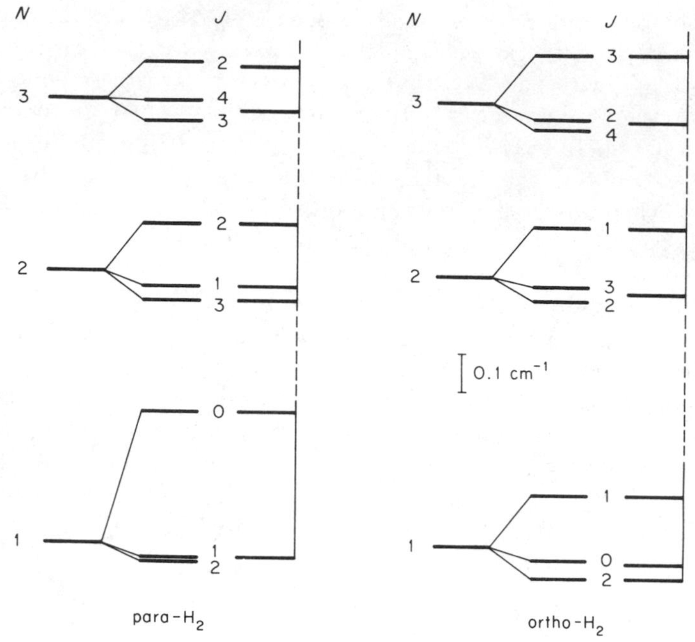

Fig. 7.1. The fine-structure splittings of the rotational levels of the lowest $^3\Pi_u$ term in H_2. The exact fit for the $N = 2$ multiplet in para-H_2, taken with the assumption that the outer electron of the pair moves in an effective spherical potential, serves to predict unambiguously all the multiplets. The calculations are presented here for $N < 4$, and compared with the partially resolved splittings found optically. The latter stand to the right of a column of J values. For the $N = 1$ level of ortho-H_2, the observed splitting is taken from the resonance experiment of Brooks *et al.* [105].

electron) and $Z = 1$ (for the 2p electron). However, we can gain a certain amount of information merely by using the hypothesis that the 2p electron is moving in a spherically symmetric potential. In the limit $r \to 0$, we have

$$|\nu, \Lambda = \pm 1\rangle \to |2\text{p}, m_l = \pm 1\rangle.$$

From the WE theorem,

$$\frac{B_0}{B_2} = \begin{pmatrix} 1 & 2 & 1 \\ -1 & 0 & 1 \end{pmatrix} \begin{pmatrix} 1 & 2 & 1 \\ -1 & 2 & -1 \end{pmatrix}^{-1} = \sqrt{\frac{1}{6}},$$

and so $B_2 = (6)^{1/2} B_0$. The formulas of Eqs. (7.20) now depend on only two parameters, A and B_0. Thus the experimental data for a given NS multiplet can be used to predict the splittings for every multiplet that derives from the same electronic and vibrational state. Lichten's data for the $N = 2$ level of $^3\Pi_u$ of para-H_2 lead to $A = -0.127$ cm^{-1}, $B_0 = -0.0591$ cm^{-1}; and these parameters have been used to calculate the multiplet splittings for $N < 4$. The results are shown in Fig. 7.1. Foster and Richardson [106] remark that their inability to completely resolve the triplets by optical methods indicates that two of the three components of a triplet lie close together, and this is borne out by the calculations. The general agreement is seen to be very good.

The united atom approximation can be carried further. As previously mentioned, we can use hydrogenic orbits for the 1s and 2p electrons of HeI 1s2p. It is now a good approximation to replace r_{12}^{-3} by the p-electron operator r^{-3}, and so we can write

$$\begin{aligned} B_0 &= 4\beta^2 \langle 2\mathrm{p} \mid r^{-3} \mid 2\mathrm{p} \rangle \langle \mathrm{p}, m_l = 1 \mid C_0^{(2)} \mid \mathrm{p}, m_l = 1 \rangle \\ &= 4\beta^2 \frac{1}{24a_0^3} \begin{pmatrix} 1 & 2 & 1 \\ -1 & 0 & 1 \end{pmatrix} (-3) \begin{pmatrix} 1 & 2 & 1 \\ 0 & 0 & 0 \end{pmatrix} \\ &= -\frac{1}{30}\beta^2 a_0^{-3} = -\frac{1}{120}\alpha^2 \epsilon_0 = -0.097 \text{ cm}^{-1}. \end{aligned} \tag{7.21}$$

If we neglect the last term on the right in Eq. (7.4), then

$$\begin{aligned} A &= -\beta^2 \langle 2\mathrm{p} \mid r^{-3} \mid 2\mathrm{p} \rangle \langle \mathrm{p}, m_l = 1 \mid l_\zeta \mid \mathrm{p}, m_l = 1 \rangle \\ &= -\tfrac{1}{24}\beta^2 a_0^{-3} = -0.121 \text{ cm}^{-1}. \end{aligned} \tag{7.22}$$

Both B_0 and A are of the same order of magnitude as the experimental values (-0.059 and -0.127) and of the same sign. The agreement in the case of A is fortuitous (see Problems 7.2 and 7.3). The united-atom approximation is thus useful for relating certain parameters (such as B_0 and B_2) and for giving rough estimates of them, but fails to provide an adequate description of their numerical values. This is scarcely surprising, of course. The reader is referred to the literature [109, 112] for more detailed theoretical treatments of A, B_0 and B_2.

7.6 THE SPIN–ROTATION INTERACTION

We have yet to examine the part H_{sr} of the Hamiltonian for H_2 that represents the interaction between the electrons' spins and the rotation of the nuclear frame. The definition of H_{sr} has already been given in Eq. (7.2). Since β_N is some three orders of magnitude smaller than β, we would expect H_{sr} to be very much less significant than H_{so}; however, its analysis is not so straightforward as that for H_{so}, and this alone makes it worthwhile to study H_{sr} in some detail. It is convenient to suppose that we are again interested only in diagonal matrix elements and that $S = 1$. We can therefore write $\mathbf{s}_t = \frac{1}{2}\mathbf{S}$, and

$$\langle[\nu\Lambda][v][N-\Lambda]SJ \mid H_{\text{sr}} \mid [\nu\Lambda][v][N-\Lambda]SJ\rangle$$
$$= 2\beta\beta_N(-1)^{N+S+J}\begin{Bmatrix} N & 1 & N \\ S & J & S \end{Bmatrix}(S \,\|\, S \,\|\, S)\,\Xi, \qquad (7.23)$$

where

$$\Xi = ([\nu\Lambda][v][N-\Lambda] \,|\!:\! \sum_{tX} r_{Xt}^{-3}(\mathbf{r}_{Xt} \times \mathbf{p}_X) \,:\!|\, [\nu\Lambda][v][N-\Lambda]). \qquad (7.24)$$

Equation (7.23) specifies the dependence of H_{sr} on J, which is identical to that of H_{so}. To find the N dependence, we need to take the operator $r_{Xt}^{-3}(\mathbf{r}_{Xt} \times \mathbf{p}_X)$ and separate the parts that are functions of the Euler angles $(\phi\theta\gamma)$ from those that are functions of the electrons' coordinates $\boldsymbol{\varrho}_t$ in the molecular frame F'. In the process of doing this, we must be ready to take note of any terms whose dependence on N is the same as that of the spin-orbit interaction H_{so}. Owing to the disparity between β_N and β, such terms can be immediately set aside and implicitly incorporated into H_{so}. They can be recognized by noting that any contribution $\mathbf{V}$ to $r_{Xt}^{-3}(\mathbf{r}_{Xt} \times \mathbf{p}_X)$ which possesses the property that $\mathbf{V}^{(01)}$ is independent of the Euler angles $(\phi\theta\gamma)$ is of precisely the same character as $\mathbf{U}$ of Section 7.2–3, and can thus be incorporated into H_{so}. Two sources for $\mathbf{V}$ are quickly apparent:

(i) As a first step to expressing $\mathbf{r}_{Xt} \times \mathbf{p}_X$ in a tractable form, we ignore translational effects and use Eqs. (6.5) to write

$$\mathbf{p}_X = -\{M_X/(M_X + M_Y)\} \sum_v \mathbf{p}_{vC} - \mathbf{p}_{YX} \qquad (X \neq Y)$$

together with $\mathbf{p}_v = \mathbf{p}_{vC}$. All tensors $(\mathbf{r}_{Xt} \times \mathbf{p}_v)^{(01)}$ are of the type $\mathbf{V}^{(01)}$ and can thus be absorbed into H_{so}. Alternatively, we may simply ignore them. This implies

$$\mathbf{p}_A = -\mathbf{p}_{BA}, \qquad \mathbf{p}_B = -\mathbf{p}_{AB} = \mathbf{p}_{BA}, \qquad (7.25)$$

and so amounts to supposing that the center of gravity C of the nuclear frame coincides with that of the entire molecule.

(ii) A second source arises when we express $\mathbf{p}_{BA}$ in spherical polars. As an incidental point, it is to be noted that

$$\langle v \mid (\partial/\partial r_{BA}) \mid v\rangle = 0,$$

since $\partial/\partial r_{BA}$ is odd with respect to inversion about the equilibrium separation r_e. This means that $\mathbf{r}_{BA} \cdot \mathbf{p}_{BA}$ is effectively zero, and so we can write (putting $r_{BA} = r$ as before)

$$\mathbf{p}_{BA} = -r^{-2}\{\mathbf{r}_{BA} \times (\mathbf{r}_{BA} \times \mathbf{p}_{BA})\} = i(2)^{1/2}r^{-1}(\mathbf{C}_{BA}{}^{(1)}\mathbf{l}_{BA}{}^{(1)})^{(1)}, \quad (7.26)$$

the last expression being derived with the aid of Eq. (1.37). The approximation of Eqs. (7.25) can now be used to obtain

$$r_{Bt}{}^{-3}(\mathbf{r}_{Bt} \times \mathbf{p}_B) = 2r^{-1}r_{Bt}{}^{-3}\{\mathbf{r}_{Bt}{}^{(1)}(\mathbf{C}_{BA}{}^{(1)}\mathbf{l}_{BA}{}^{(1)})^{(1)}\}^{(1)},$$
$$r_{At}{}^{-3}(\mathbf{r}_{At} \times \mathbf{p}_A) = -2r^{-1}r_{At}{}^{-3}\{\mathbf{r}_{At}{}^{(1)}(\mathbf{C}_{BA}{}^{(1)}\mathbf{l}_{BA}{}^{(1)})^{(1)}\}^{(1)}. \quad (7.27)$$

As soon as we write $\mathbf{l}_{BA} = \mathbf{N} - \mathbf{L}$ (which follows from Eqs. (6.25)–(6.26)), a source for $\mathbf{V}$ is provided by the part $-\mathbf{L}$. This is because $\mathbf{L}^{(01)}$ is independent of the Euler angles. Thus the terms

$$2r^{-1}r_{Xt}{}^{-3}\{\mathbf{r}_{Xt}{}^{(1)}(\mathbf{C}_{BA}{}^{(1)}\mathbf{L}^{(1)})\}^{(1)} \quad (7.28)$$

contribute to H_{so} and are therefore dropped. The vector $\mathbf{N}^{(01)}$, however, is a function of the Euler angles (see Eq. (6.21)). We are left with the modified Hamiltonian

$$H_{\text{sr}}' = 2\beta\beta_N \sum_{Xt} \mathbf{T}_{Xt}\cdot\mathbf{S}, \quad (7.29)$$

where

$$\mathbf{T}_{At} = -2r^{-1}r_{At}{}^{-3}\{\mathbf{r}_{At}{}^{(1)}(\mathbf{C}_{BA}{}^{(1)}\mathbf{N}^{(1)})^{(1)}\}^{(1)},$$
$$\mathbf{T}_{Bt} = 2r^{-1}r_{Bt}{}^{-3}\{\mathbf{r}_{Bt}{}^{(1)}(\mathbf{C}_{BA}{}^{(1)}\mathbf{N}^{(1)})^{(1)}\}^{(1)}.$$

The next step is to use the result of Problem 1.4 to give

$$\mathbf{C}_{BA}{}^{(1)} = -(3)^{-1/2}\mathbf{D}_{\cdot 0}{}^{(11)}(\phi\theta\gamma). \quad (7.30)$$

The null projection in F' can be transferred to the outermost part of the coupled tensor product by assigning a second rank to the vectors. Thus we get

$$\mathbf{T}_{Bt} = -2(3)^{-1/2}r^{-1}r_{Bt}{}^{-3}\{\mathbf{r}_{Bt}{}^{(10)}(\mathbf{D}^{(11)}\mathbf{N}^{(10)})^{(11)}\}_{\cdot 0}{}^{(11)}. \quad (7.31)$$

Since $\mathbf{r}_{Bt}$ refers to the electrons' coordinates, we write

$$\mathbf{r}_{Bt}{}^{(10)} = (\mathbf{D}^{(11)}\mathbf{r}_{Bt}{}^{(01)})^{(10)},$$

thereby separating the part depending on $(\phi\theta\gamma)$ from that depending on $(r\varrho_t)$. A recoupling is now performed to contract the two tensors $\mathbf{D}^{(11)}$. We need the two coefficients

$$((10)1,\ (11)1,\ 1\,|\,(11)k,\ (01)1,\ 1) = \{3(2k+1)\}^{1/2}\begin{Bmatrix}1 & 1 & k\\ 1 & 1 & 1\end{Bmatrix},$$

$$((11)0,\ (10)1,\ 1\,|\,(11)k,\ (10)1,\ 1) = \tfrac{1}{3}(2k+1)^{1/2},$$

and also, from Eq. (1.42), the result

$$(\mathbf{D}^{(11)}\mathbf{D}^{(11)})^{(kk)} = 3(2k+1)^{-1/2}\mathbf{D}^{(kk)}.$$

Combining the various parts, we get

$$\mathbf{T}_{Bt}{}^{(11)} = -2r^{-1}\sum_k (2k+1)^{1/2}\begin{Bmatrix}1 & 1 & k\\ 1 & 1 & 1\end{Bmatrix} r_{Bt}{}^{-3}\{\mathbf{D}^{(kk)}(\mathbf{r}_{Bt}{}^{(01)}\mathbf{N}^{(10)})^{(11)}\}^{(11)}.$$

Ranks are attached to $\mathbf{T}_{Bt}$ to match those on the right-hand side of the equation. A further recoupling, for which

$$(k, (10)1, 1 \mid (k1)1, 0, 1) = (k, (01)1, 1 \mid (k0)k, 1, 1) = (-1)^k,$$

brings $\mathbf{D}^{(kk)}$ and $\mathbf{N}^{(10)}$ together, thus completing the coordinate separation:

$$\mathbf{T}_{Bt}{}^{(11)} = -2r^{-1}\sum_k (2k+1)^{1/2}\begin{Bmatrix}1 & 1 & k\\ 1 & 1 & 1\end{Bmatrix} r_{Bt}{}^{-3}\{(\mathbf{D}^{(kk)}\mathbf{N}^{(10)})^{(1k)}\mathbf{r}_{Bt}{}^{(01)}\}^{(11)}. \tag{7.32}$$

The coordinates $(\phi\theta\gamma)$ appear only in $(\mathbf{D}^{(kk)}\mathbf{N}^{(10)})^{(1k)}$. For this tensor, we rapidly obtain

$$(N \,\|\, (\mathbf{D}^{(kk)}\mathbf{N}^{(10)})^{(1k)} \,\|\, N)$$
$$= -\{3(2k+1)N(N+1)(2N+1)^3\}^{1/2}\begin{Bmatrix}1 & 1 & k\\ N & N & N\end{Bmatrix} \tag{7.33}$$

from the result of Problem 1.6 (extended to double tensors). For a given Λ state, only $(r_{00}{}^{(01)})_{Bt}$ can contribute to the matrix element Ξ. We find that

$$([\nu\Lambda][v][N-\Lambda] \,|\!:\, \sum_{Xt} (T_{\cdot 0}{}^{(11)})_{Xt} \,:\!|\, [\nu\Lambda][v][N-\Lambda]) = \Theta W,$$

where

$$W = ([\nu\Lambda][v] \mid r^{-1}\sum_t \{r_{Bt}{}^{-3}(r_{00}{}^{(01)})_{Bt} - r_{At}{}^{-3}(r_{00}{}^{(01)})_{At}\} \mid [\nu\Lambda][v]), \tag{7.34}$$

and

$$\Theta = 2\sum_k (2k+1)(k0, 10 \mid 10)\{3N(N+1)(2N+1)^3\}^{1/2}$$
$$\times (-1)^{N+\Lambda}\begin{pmatrix}N & k & N\\ \Lambda & 0 & -\Lambda\end{pmatrix}\begin{Bmatrix}1 & 1 & k\\ N & N & N\end{Bmatrix}\begin{Bmatrix}1 & 1 & k\\ 1 & 1 & 1\end{Bmatrix}$$

Putting in the explicit expressions for the 3-j and 6-j symbols, we get

$$\Theta = \{N(N+1)(2N+1)\}^{1/2}\left\{1 - \frac{\Lambda^2}{N(N+1)}\right\}. \qquad (7.35)$$

Collecting our results together, we have

$$\langle[\nu\Lambda][v][N-\Lambda]SJ \mid H_{sr}' \mid [\nu\Lambda][v][N-\Lambda]SJ\rangle$$
$$= \beta\beta_N\{J(J+1) - S(S+1) - N(N+1)\}\left\{1 - \frac{\Lambda^2}{N(N+1)}\right\}W. \qquad (7.36)$$

This expresses both the J and the N dependence of the diagonal matrix elements of H_{sr}'.

For sufficiently large N, the term in Λ^2 in Eq. (7.36) can be ignored, and the overall multiplet splitting becomes proportional to N itself. This is in contrast to H_{so}, for which the splittings are proportional to N^{-1}. Of course, the term in Λ^2 could be merged with H_{so} if we so wished. A precisely similar equation to (7.36) can be written down for any state for which the replacements of the type $\mathbf{s}_i \to \kappa\mathbf{S}$ can be made: we only have to multiply the right-hand side by 2κ.

It might also be remarked that the united-atom approximation cannot be expected to work well in estimating the effect of H_{sr}' on the multiplet splittings of the $^3\Pi$ term of H_2. The major role is played by the orbital that goes over into the 1s eigenfunction of He^+ in the limit $r \to 0$. This molecular orbital presumably resembles the lowest orbital of H_2^+ quite closely [114] and it cannot be adequately described within a central-field model.

7.7 EFFECTIVE OPERATORS

We have already remarked in Section 7.3 that the spin-orbit splittings within a single multiplet can be reproduced by $\lambda\mathbf{N}\cdot\mathbf{S}$. This operator gives the correct J dependence, but fails to reproduce the dependence on N (as exhibited in Eq. (7.8)). It is natural to ask whether other operators might not be more successful; and, furthermore, whether one could be found to reproduce both the N and J dependences of the matrix elements of the spin-rotation interaction H_{sr}'.

An important criterion is that the matrix elements of an effective operator should be easy to evaluate. We want to avoid factors such as W of Eq. (7.34). This considerably limits our choice. Scalar products of the type $\mathbf{L}\cdot\mathbf{S}$ or $\mathbf{R}\cdot\mathbf{S}$ suggest themselves, for not only are their matrix elements easy to calculate, but we can also be sure that the J dependence is correctly reproduced for both H_{so} and H_{sr}'. In fact, for $\mathbf{L}\cdot\mathbf{S}$, we have simply to make the substitution $U_{00}^{(01)} \to L_{00}^{(01)}$ in Eq. (7.8). The eigenvalue of $L_{00}^{(01)}$ is Λ, and so

$$\langle (NS)J \mid \mathbf{L}\cdot\mathbf{S} \mid (NS)J \rangle = \Lambda^2 \frac{J(J+1) - N(N+1) - S(S+1)}{2N(N+1)}. \tag{7.37}$$

Thus, for the spin–orbit interaction, we can write

$$H_{so} \equiv \lambda' \mathbf{L}\cdot\mathbf{S},$$

where λ' is independent of J and N.

The dependence of the right-hand side of Eq. (7.37) on J and N is identical to that of the Λ^2 term in Eq. (7.36). If we combine that result with

$$\langle (NS)J \mid \mathbf{N}\cdot\mathbf{S} \mid (NS)J \rangle = \tfrac{1}{2}\{J(J+1) - N(N+1) - S(S+1)\}$$

and use the equation $\mathbf{R} = \mathbf{N} - \mathbf{L}$, we obtain the equivalence

$$H_{sr}' \equiv \Gamma \mathbf{R}\cdot\mathbf{S}, \tag{7.38}$$

where $\Gamma = 2\beta\beta_N W$. The success of $\mathbf{L}\cdot\mathbf{S}$ and $\mathbf{R}\cdot\mathbf{S}$ in reproducing the spin–orbit and spin–rotation interactions accords with what we might intuitively have expected from the nature of $\mathbf{L}$, $\mathbf{R}$, and $\mathbf{S}$.

It is worth noting that Eq. (7.36) can be checked by making the artificial hypothesis that $r_{At}^{-3} = r_{Bt}^{-3}$. Although physically quite unrealistic, this assumption allows us to write

$$H_{sr} = 4\beta\beta_N \sum_t r_{Xt}^{-3}\{(\mathbf{r}_{At} \times \mathbf{p}_A) + (\mathbf{r}_{Bt} \times \mathbf{p}_B)\}\cdot\mathbf{s}_t.$$

With the aid of the approximate equations (7.25), we have

$$\begin{aligned}(\mathbf{r}_{At} \times \mathbf{p}_A) + (\mathbf{r}_{Bt} \times \mathbf{p}_B) &= (\mathbf{r}_{AB} + \mathbf{r}_{Bt}) \times \mathbf{p}_{AB} - (\mathbf{r}_{Bt} \times \mathbf{p}_{AB}) \\ &= \mathbf{l}_{AB} = \mathbf{l}_{BA} = \mathbf{R}.\end{aligned}$$

The approximations (7.25) were also used in deriving H_{sr}'; and, furthermore, we can show that the discarded terms (7.28) contribute nothing when $r_{At}^{-3} = r_{Bt}^{-3}$ (see Problem 7.4). So, under these conditions,

$$H_{sr}' = 2\beta\beta_N \mathbf{R}\cdot\mathbf{S} \sum_t r_{Xt}^{-3}.$$

It remains for us to confirm that

$$W \equiv \sum_t r_{Xt}^{-3}.$$

This we can do very simply, for, under the condition $r_{At}^{-3} = r_{Bt}^{-3}$,

$$r^{-1} \sum_t \{r_{Bt}^{-3}(r_{00}^{(01)})_{Bt} - r_{At}^{-3}(r_{00}^{(01)})_{At}\} = r^{-1} \sum_t r_{Xt}^{-3}(r_{00}^{(01)})_{BA}$$

$$= \sum_t r_{Xt}^{-3}.$$

This provides a very satisfying check on the analysis of Section 7.6.

7.8 HYPERFINE INTERACTION

It is particularly simple to study the hyperfine interaction for H_2 because of the symmetry between the two electrons and the two protons. To get the contributions to the Hamiltonian that are linear in the nuclear spins $\mathbf{I}_A$ and $\mathbf{I}_B$, we take Eqs. (7.1) and (7.2) and make the replacements

$$t \leftrightarrow X, \qquad -2\beta\mathbf{s} \rightarrow \beta_N\mu_N\mathbf{I}/I, \qquad \beta\mathbf{p}_t \leftrightarrow -\beta_N\mathbf{p}_X.$$

These substitutions interchange particles, magnetic moments, and currents. Thus μ_N is the magnetic moment of a proton measured in nuclear magnetons ($\mu_N = 2.793$). Neglecting terms in β_N^2, we obtain

$$H_{\text{no}} = -\frac{2\beta\beta_N\mu_N}{I} \sum_{Xt} \left[\boldsymbol{\nabla}_t\left(\frac{1}{r_{tX}}\right) \times \mathbf{p}_t\right]\cdot\mathbf{I}_X \tag{7.39}$$

for the interaction between the nuclear spins and the orbital motion of the electrons. Since Eqs. (7.1)–(7.2) exclude nuclear spin, H_{no} cannot be expected to include contributions from electron spin. In fact, the terms linear in both $\mathbf{s}$ and $\mathbf{I}$ are not difficult to write down. The spin–spin interaction comprises the dipolar term

$$H_{\text{ns}} = -\frac{2\beta\beta_N\mu_N}{I} \sum_{Xt} \left[\frac{\mathbf{s}_t\cdot\mathbf{I}_X}{r_{tX}^3} - \frac{3(\mathbf{r}_{tX}\cdot\mathbf{s}_t)(\mathbf{r}_{tX}\cdot\mathbf{I}_X)}{r_{tX}^5}\right], \tag{7.40}$$

which follows from Eq. (7.9); and also a contact term [115]

$$H_{\text{nsc}} = \frac{2\beta\beta_N\mu_N}{I}\frac{8\pi}{3} \sum_{Xt} \delta(\mathbf{r}_{tX})\mathbf{s}_t\cdot\mathbf{I}_X. \tag{7.41}$$

The most satisfactory way of deriving this last contribution is to examine the nonrelativistic limit of a Dirac electron moving in the magnetic potential of a dipole $\boldsymbol{\mu}$ given by

$$\boldsymbol{\mu} = \beta_N\mu_N\mathbf{I}/I. \tag{7.42}$$

Various limiting procedures can also be used to derive Eq. (7.41) (see, for example, Problem 7.5). Throughout this paragraph we could, of course, set $I = \frac{1}{2}$; but it is convenient not to do so since Eqs. (7.39)–(7.41) describe the hyperfine interaction (linear in the nuclear spins) for any diatomic molecule comprising two identical nuclei with nuclear moments μ_N. The only simplification afforded by having $I = \frac{1}{2}$ is the absence of quadrupole interactions. The general case was first studied by Frosch and Foley [116]. A more recent review of hyperfine structures of diatomic molecules has been given by Dunn [117].

For H_2 we cannot always suppose that the hyperfine interaction is appreciably weaker than fine-structure effects. If, nevertheless, we restrict our attention to cases where this hypothesis is not too gross an approximation, the molecular states are conveniently written as

$$|\,[\nu\Lambda][v][N-\Lambda]SJ, TF\rangle, \tag{7.43}$$

where $\mathbf{J}$ is coupled to the total nuclear spin $\mathbf{T}$ (the resultant of $\mathbf{I}_A$ and $\mathbf{I}_B$) to produce a total angular momentum vector $\mathbf{F}$. We may hope to obtain a reasonable representation of the effects of hyperfine interactions by calculating the diagonal matrix elements of H_{no}, H_{ns}, and H_{nsc} in the scheme (7.43).

As with the analyses of Section 7.3–7.6, our primary aim is to separate out the dependence on as many angular momentum vectors as possible. The process is rather more straightforward than that for H_{sr}, and we simply give the results:

$$\begin{aligned}\langle[\nu\Lambda][v][N-\Lambda]SJTF\,|\,H_{\text{no}}\,|\,[\nu\Lambda][v][N-\Lambda]SJTF\rangle \\ = 2a[J,T,N](-1)^{2I+F-N-S}\sqrt{\frac{I(I+1)(2I+1)}{N(N+1)(2N+1)}} \\ \times\begin{Bmatrix}J & 1 & J\\ T & F & T\end{Bmatrix}\begin{Bmatrix}J & 1 & J\\ N & S & N\end{Bmatrix}\begin{Bmatrix}T & 1 & T\\ I & I & I\end{Bmatrix},\end{aligned} \tag{7.44}$$

where

$$a = \Lambda(\beta\beta_N\mu_N/I)\langle[\nu\Lambda][v]\,|\sum_{Xt} r_{tX}^{-3}(\lambda_\zeta)_t\,|\,[\nu\Lambda][v]\rangle;$$

and, for a two-electron system for which $S = 1$,

$$\begin{aligned}\langle[\nu\Lambda][v][N-\Lambda]SJTF\,|\,H_{\text{nsc}}\,|\,[\nu\Lambda][v][N-\Lambda]SJTF\rangle \\ = 4b[J,T,S](-1)^{2I+F-S-N}\sqrt{\frac{I(I+1)(2I+1)}{S(S+1)(2S+1)}} \\ \times\begin{Bmatrix}J & 1 & J\\ T & F & T\end{Bmatrix}\begin{Bmatrix}J & 1 & J\\ S & N & S\end{Bmatrix}\begin{Bmatrix}T & 1 & T\\ I & I & I\end{Bmatrix},\end{aligned} \tag{7.45}$$

where

$$b = (4\pi\beta\beta_N\mu_N/3I)\langle[\nu\Lambda][v] \mid \sum_{Xt} \delta(\mathbf{r}_{tX}) \mid [\nu\Lambda][v]\rangle.$$

Equations (7.44) and (7.45) specify the dependences of the diagonal matrix elements on J, T, N, and F. They also expose an unexpected symmetry between N and S. We can deduce that the quantities a and b must enter through the single linear combination $a + 2b$ for the $N = 1$ level of the $^3\Pi_u$ state of ortho-H_2.

The analysis is only slightly more complicated for H_{ns}. Like H_{ss}, this operator can connect states differing by two units in Λ. We begin by replacing Eq. (7.11) by

$$H_{\text{ns}} = 2(6)^{1/2}(\beta\beta_N\mu_N/I) \sum_{Xt} r_{tX}^{-3}(\mathbf{C}^{(2)}(\theta_{tX}\phi_{tX}) \cdot (\mathbf{s}_t\mathbf{I}_X)^{(2)}). \qquad (7.46)$$

To match the coupling represented in (7.43), we make the conversion

$$\mathbf{C}^{(2)} \cdot (\mathbf{s}_t\mathbf{I}_X)^{(2)} = -(5/3)^{1/2}(\mathbf{C}^{(2)}\mathbf{s}_t)^{(1)} \cdot \mathbf{I}_X^{(1)},$$

thereby obtaining, for a two-electron system with $S = 1$,

$$\langle[\nu\Lambda][v][N - \Lambda]SJTF \mid H_{\text{ns}} \mid [\nu\Lambda'][v][N - \Lambda']SJTF\rangle$$

$$= (120)^{1/2} e_{|\Lambda-\Lambda'|}[J, T, N, S](-1)^{2I+J+F+N+\Lambda'} \sqrt{\frac{I(I+1)(2I+1)}{S(S+1)(2S+1)}}$$

$$\times \begin{pmatrix} N & 2 & N \\ \Lambda & \Lambda' - \Lambda & -\Lambda' \end{pmatrix} \begin{Bmatrix} J & 1 & J \\ T & F & T \end{Bmatrix} \begin{Bmatrix} T & 1 & T \\ I & I & I \end{Bmatrix} \begin{Bmatrix} N & N & 2 \\ S & S & 1 \\ J & J & 1 \end{Bmatrix}, \quad . \qquad (7.47)$$

where

$$e_{|\Lambda-\Lambda'|} = (\beta\beta_N\mu_N/I)\langle[\nu\Lambda][v] \mid \sum_{Xt} r_{tX}^{-3} C_{\Lambda-\Lambda'}^{(2)}(\vartheta_{tX}\varphi_{tX}) \mid [\nu\Lambda'][v]\rangle.$$

To produce the notation of Chiu [110] we write $e_0 = c$ and $e_2 = d$. (In the notation of Jette and Cahill [111] $e_0 = \frac{2}{3}c$ and $e_2 = (5/3)^{1/2}d$, while $b = a_F$.)

We can now assemble our results for the $N = 1$ level of $^3\Pi_u$ of ortho-H_2. Putting $I = \frac{1}{2}$, $S = 1$, $N = 1$, and $T = 1$ in Eqs. (7.44)–(7.47), the diagonal energies E_{JF} of the total hyperfine interaction (linear in $\mathbf{I}$) become

$$E_{JF} = (1/80)[F(F+1) - J(J+1) - 2][10(a + 2b) + \{3J(J+1) - 16\}(c - d\sqrt{6})]. \qquad (7.48)$$

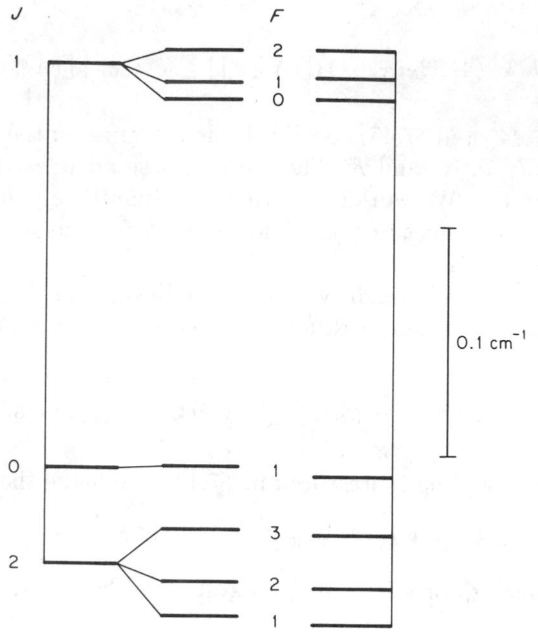

Fig. 7.2. Hyperfine structure of the $N = 1$ level of the lowest $^3\Pi_u$ state of ortho-H_2. On the left are shown the theoretical levels and their origins in the $N = 1$ triplet. Only diagonal matrix elements are used in the calculation. The experimental levels of Brooks *et al.* [105] are given on the right. The structure is repeated, with small variations, for each vibrational level v.

A good fit [105] with the experimental data can be obtained by taking

$$a + 2b = 0.0309 \text{ cm}^{-1}, \qquad c - d\sqrt{6} = 0.0023 \text{ cm}^{-1}.$$

This is illustrated in Fig. 7.2. The fit can be improved by considering off-diagonal matrix elements within the $N = 1$ manifold, as has been done by Brooks *et al.* [105]. The parameters a and b can now be separated, and it is found that b is 0.0150 cm^{-1}. The hyperfine structure is thus dominated by the Fermi contact terms. The value of b is very close to 0.0147 cm^{-1}, which Jefferts [118] has obtained (by extrapolation) for the $(Nv) \equiv (10)$ level of the ground state $^2\Sigma$ of ortho-H_2^+. This result can be regarded as providing direct support for McDonald's model of the excited state $^3\Pi_u$ of H_2 (see Section 7.6).

7.9 THE GROUND STATE $^1\Sigma$

Small interactions involving nuclear spins can be exhibited by studying the electronically inert ground state $^1\Sigma$ of ortho-H_2. A good account has been given by Ramsey [119]. To complete our examination of the interparticle interactions in H_2, we must consider the dipolar interaction H_{nn} between nuclear spins, and the small terms (involving β_N^2) which were dropped in writing down Eq. (7.40). These last correspond to the interaction H_{nr} between the nuclei and the rotation of the molecule.

In complete analogy to H_{ss}, we have

$$H_{nn} = \left(\frac{\beta_N\mu_N}{I}\right)^2\left[\frac{\mathbf{I}_A\cdot\mathbf{I}_B}{r_{AB}^3} - \frac{3(\mathbf{r}_{AB}\cdot\mathbf{I}_A)(\mathbf{r}_{AB}\cdot\mathbf{I}_B)}{r_{AB}^5}\right]$$

$$= -(6)^{1/2}(\beta_N\mu_N/I)^2 r_{AB}^{-3}(\mathbf{C}^{(2)}(\theta_{AB}\phi_{AB})\cdot(\mathbf{I}_A\mathbf{I}_B)^{(2)}).$$

This is quite straightforward to cope with. We use

$$\mathbf{C}^{(2)}(\theta_{AB}\phi_{AB}) = \mathbf{C}^{(2)}(\theta\phi) = (5)^{-1/2}\mathbf{D}_{\cdot 0}{}^{(22)}(\phi\theta\gamma)$$

to obtain, for $T = 1$,

$$\langle[{}^1\Sigma][v][N0]TF \mid H_{nn} \mid [{}^1\Sigma][v][N0]TF\rangle$$

$$= (\beta_N\mu_N/I)^2(-1)^{N+F+1}\begin{Bmatrix} N & 2 & N \\ 1 & F & 1 \end{Bmatrix}\sqrt{\frac{15N(N+1)(2N+1)}{2(2N-1)(2N+3)}}$$

$$\times \langle[{}^1\Sigma][v] \mid r_{AB}^{-3} \mid [{}^1\Sigma][v]\rangle. \qquad (7.49)$$

The experimental analysis of the $N = 1$, $v = 0$ level of the $^1\Sigma$ ground state of H_2 yields a value for $\langle r_{AB}^{-3}\rangle$.

The interaction H_{nr} leads to a different F dependence from that contained in the 6-*j* symbol of Eq. (7.49). The analysis of Section 7.2 that reduced H_{so} to the effective operator (7.3) can be repeated for H_{nr}, since the proton–electron interchange holds good when the condition $S = 1$ is replaced by $T = 1$. Thus, in analogy to Eq. (7.3), we write

$$H_{nr} = (\mu_N\beta_N^2/2I)\,\mathbf{V}\cdot\mathbf{T},$$

where $\mathbf{V}$ parallels the vector $\mathbf{U}$ defined in Eq. (7.4). Carrying the correspondence further, we take

$$H_{nr} \equiv \lambda''\mathbf{N}\cdot\mathbf{T}$$

for a manifold of states of constant N. This effective operator has the matrix elements

$$\tfrac{1}{2}\lambda''\{F(F+1) - N(N+1) - T(T+1)\},$$

whereas the matrix elements of H_{nn}, given in Eq. (7.49), are quadratic in $F(F+1)$.

The experimental values [119] for the level of $^1\Sigma$ of ortho-H_2 for which $(N, v) \equiv (1, 0)$ are

$$(\beta_N\mu_N/I)^2\langle r_{AB}^{-3}\rangle = 96 \times 10^{-7}\ \mathrm{cm}^{-1}, \qquad \lambda'' = -38 \times 10^{-7}\ \mathrm{cm}^{-1}.$$

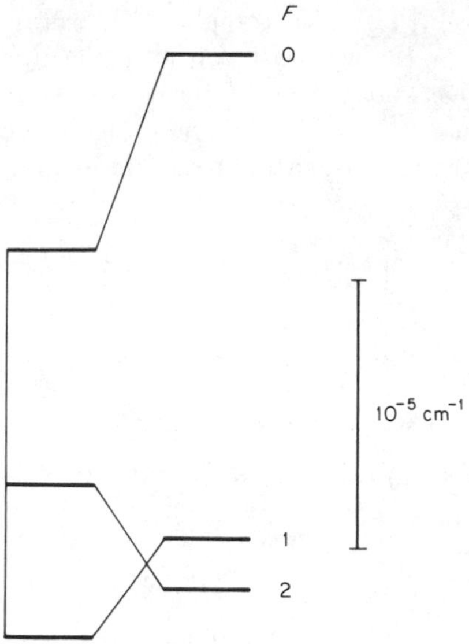

Fig. 7.3. Hyperfine structure of the $N = 1$, $v = 0$ level of the $^1\Sigma$ ground state of ortho-H_2. The splitting on the left corresponds to the dipolar interaction H_{nn} between the two protons. To this structure are added the contributions from the interaction H_{nr} of the nuclear spins with the molecular rotation. The strengths of H_{nn} and H_{nr} are chosen to give an exact fit with experiment.

The hyperfine multiplet is constructed in Fig. 7.3 to show the relative importance of H_{nn} and H_{nr}. It is apparent from the scale that the splittings are very much smaller than those shown in Figs. 7.1 and 7.2.

PROBLEMS

7.1 Equations (7.20) give the energies of the three components that comprise a fine-structure multiplet occurring in the rotational levels of $^3\Pi_u$ of H_2. Prove that, in the limit of large N, the two components for which $J = N \pm 1$ tend to coalesce. Show also (a) that the interval between this coalesced level and the level for which $J = N$ is determined solely by the spin-spin interaction; and (b) that these two levels correspond to co-directional spins (for $J = N \pm 1$) and opposed spins (for $J = N$).

7.2 In the approximation for HeI 1s2p for which the 1s and 2p electrons are represented by hydrogenic eigenfunctions corresponding to $Z = 2$ and $Z = 1$ respectively, prove that

$$\frac{((\mathrm{sp})\ {}^3\mathrm{P} \,\|\, (\mathbf{r}_2 \times \mathbf{p}_1 + \mathbf{r}_1 \times \mathbf{p}_2) \,\|\, (\mathrm{sp})\ {}^3\mathrm{P})}{((\mathrm{sp})\ {}^3\mathrm{P} \,\|\, (\mathbf{r}_1 \times \mathbf{p}_1 + \mathbf{r}_2 \times \mathbf{p}_2) \,\|\, (\mathrm{sp})\ {}^3\mathrm{P})} = -\frac{2^{17}}{5^9}.$$

Hence show that the value of $|A|$ given in Eq. (7.22) should be increased by approximately 20% if the complete form for $\mathbf{U}$ (as given in Eq. (7.4)) is used.

7.3 Prove that the hydrogenic approximation of the previous problem leads to a fine structure for the multiplet ^3P of He I 1s2p which is identical to the fine structure of the $N = 1$ level of the $^3\Pi_u$ state of ortho-H_2 in the united-atom approximation provided A is increased by a factor of 2 and B_0 by a factor of -2. The experimental splittings of HeI 1s2p ^3P are given, in cm^{-1}, by [113]

$$E(^3P_0) - E(^3P_1) = 0.99, \qquad E(^3P_1) - E(^3P_2) = 0.08.$$

Use these experimental results together with the factors ± 2 to yield directly from experiment the united-atom parameters

$$A = -0.197, \qquad B_0 = -0.106$$

for the NS multiplets deriving from the $^3\Pi_u$ level of H_2. (These figures are to be compared with the calculated ones of Eqs. (7.21)–(7.22.)

7.4 The operator $\mathbf{V}_X{}^{(01)}$ is defined by

$$\mathbf{V}_X{}^{(01)} = 2r^{-1}r_{Xt}{}^{-3}\{\mathbf{r}_{Xt}{}^{(01)}(\mathbf{C}_{BA}{}^{(01)}\mathbf{L}^{(01)})^{(01)}\}^{(01)}.$$

Show that, of the three components L_ξ, L_η and L_ζ of $\mathbf{L}^{(01)}$, only L_ξ and L_η survive when the coupled parts of the above expression are expanded. Prove

$$(V_{00}{}^{(01)})_A - (V_{00}{}^{(01)})_B = r^{-1}(r_{Bt}{}^{-3} - r_{At}{}^{-3})(\xi_t L_\xi + \eta_t L_\eta),$$

where

$$\xi_t = \xi_{At} = \xi_{Bt}, \qquad \eta_t = \eta_{At} = \eta_{Bt}.$$

Deduce that $H_{sr} = H_{sr}'$ (in the notation of Section 7.6) if we set $r_{Bt}{}^{-3} = r_{At}{}^{-3}$.

7.5 In the limit $r \to 0$, the spin–rotation interaction H_{sr}' goes over into the atomic hyperfine interaction for which the nuclear moment $\boldsymbol{\mu}$ is $\beta_N\mathbf{N}$. Prove that, in the notation of Section 7.5,

$$r^{-1}\{r_{Bt}{}^{-3}(r_{00}{}^{(01)})_{Bt} - r_{At}{}^{-3}(r_{00}{}^{(01)})_{At}\} \to -2\rho_t{}^{-3}C_0{}^{(2)}(\vartheta_t\varphi_t) + (4\pi/3)\delta(\boldsymbol{\varrho}_t)$$

in the limit $r \to 0$, where $(\rho_t\vartheta_t\varphi_t)$ are the polar coordinates of electron t in the molecular frame F'. Hence show from Eq. (7.36) that the atomic hyperfine structure coming from 1s2s $^3S_1$ of He I is given by

$$\frac{4\pi}{3}\beta\beta_N\{F(F+1) - S(S+1) - I(I+1)\}\{|\psi_{1s}(0)|^2 + |\psi_{2s}(0)|^2\},$$

where $\psi_{ns}(0)$ is the magnitude of the ns eigenfunction at the coalesced nucleus, and where the quantum numbers F and I appear in the limits $J \to F$, $N \to I$. Verify that an identical expression arises from the Fermi hyperfine contact term [115]

$$\frac{16\pi}{3}\beta \sum_t \delta(\boldsymbol{\varrho}_t)(\mathbf{s}_t \cdot \boldsymbol{\mu}).$$

7.6 Show that, for a 1s electron in He^+,

$$|\psi(0)|^2 = 8/\pi a_0^3.$$

If the spin-rotation interaction for the $^3\Pi_u$ state of H_2 is assumed to arise entirely from the orbital which, in the united-atom approximation, becomes the 1s eigenfunction of He, show that, in this approximation, the contribution of H_{sr}' to the parameter A of Eqs. (7.8) and (7.19) is

$$(64\beta\beta_N/3a_0^3)(N^2 + N - 1).$$

Jette [113a] has used $C(N^2 + N - 1)$ to fit the experimental data, where $C = 0.0008\ \text{cm}^{-1}$. Show that the united-atom model overestimates this coefficient by more than an order of magnitude.

7.7 Use an explicit expression for the 6-j symbol in Eq. (7.49) to show that, within a manifold of states of constant N and T, the interaction H_{nn} between nuclear spins for a $^1\Sigma$ state of ortho-H_2 can be represented by the effective operator

$$H_{\text{nn}} \equiv \left(\frac{\beta_N \mu_N}{I}\right)^2 \langle r_{AB}^{-3}\rangle \frac{1}{(2N-1)(2N+3)} \{3(\mathbf{T}\cdot\mathbf{N})^2 + \frac{3}{2}(\mathbf{T}\cdot\mathbf{N}) - \mathbf{T}^2\mathbf{N}^2\}$$

(see Ramsey [119]).

7.8 The coupling for the states $^2\Sigma$ of ortho-H_2^+ is rather well represented by the kets $|(ST)F_2NF\rangle$, in which the total nuclear spin T $(=1)$ is coupled to the electronic spin to produce the intermediate angular momentum F_2 [118]. Show that the hyperfine interaction could be represented by the effective operator $b\mathbf{T}\cdot\mathbf{S}$ if the contact term H_{nsc} were the sole source of the interaction, but that the inclusion of the dipolar contribution H_{ns} requires the addition of the term $cT_\zeta S_\zeta$ (as well as an adjustment to the parameter b), where

$$c = 3(\beta\beta_N\mu_N/2I)\langle \sum_X r_{tX}^{-3}(3\cos^2\vartheta_{tX} - 1)\rangle.$$

Prove that the term $cT_\zeta S_\zeta$ splits a given F_2 level, the energies of the components relative to the unsplit level being (for $S = \frac{1}{2}$, $T = 1$, and $c \ll b$)

$$c(30)^{1/2}(-1)^{F_2+F}[N, F_2] \begin{pmatrix} N & 2 & N \\ 0 & 0 & 0 \end{pmatrix} \begin{Bmatrix} F_2 & 2 & F_2 \\ N & F & N \end{Bmatrix} \begin{Bmatrix} 1 & 1 & 1 \\ \frac{1}{2} & \frac{1}{2} & 1 \\ F_2 & F_2 & 2 \end{Bmatrix}.$$

Verify that the addition of a small spin–rotation interaction $d\mathbf{S}\cdot\mathbf{N}$ adds

$$d\{3N(N+1)(2N+1)/2\}^{1/2}(-1)^{N+F-1/2}[F_2] \begin{Bmatrix} F_2 & 1 & F_2 \\ N & F & N \end{Bmatrix} \begin{Bmatrix} F_2 & 1 & F_2 \\ \frac{1}{2} & 1 & \frac{1}{2} \end{Bmatrix}$$

to the above expression.

7.9 The eigenfunction of a rotating spheroidal nucleus is $|[\psi][IM\ -N]\rangle$, where $[\psi]$ is the intrinsic nuclear eigenfunction and $[IM\ -N]$ is the rotational part. Show that, for an external observer, the maximum nuclear quadrupole moment Q (corresponding to $M = I$) is related to the intrinsic quadrupole moment Q_0 by

$$Q = \frac{3N^2 - I(I+1)}{(I+1)(2I+3)} Q_0.$$

Deduce that for $N = I$ (corresponding, in practice, to the nuclear ground state),

$$Q = \frac{I(2I-1)}{(I+1)(2I+3)} Q_0$$

(see Moszkowski [120]).

7.10 If the bipolar expansion of Eq. (5.17) is used to evaluate the energy of interaction between two separated charge distributions of densities $\rho(\mathbf{r}_{1A})$ and $\rho'(\mathbf{r}_{2B})$ that lie in the vicinity of A and B, we can select the terms for which $l = k = 1$ to obtain an expression for the interaction energy of two dipoles. By taking

$$\int \rho(\mathbf{r}_{1A})\mathbf{r}_{1A}\, d\mathbf{r}_{1A} = -2\beta\mathbf{s}_1,$$

$$\int \rho'(\mathbf{r}_{2B})\mathbf{r}_{2B}\, d\mathbf{r}_{2B} = -2\beta\mathbf{s}_2,$$

and setting

$$r_{AB}^{-3}\mathbf{C}_{AB}^{(2)} \longrightarrow r_{12}^{-3}\mathbf{C}_{12}^{(2)},$$

show that the expression for the dipole–dipole interaction of two electronic spins, as given in Eq. (7.11), is obtained.

8

External Fields

8.1 INTRODUCTION

In the nonrelativistic limit, it is comparatively easy to write down the terms that must be added to the Hamiltonian of a diatomic molecule when an external electromagnetic field is present. A particle of momentum $\hbar\mathbf{p}$, charge q, mass M, and magnetic moment $\boldsymbol{\mu}$ provides a contribution to the Hamiltonian of

$$q\phi - \boldsymbol{\mu}\cdot\mathbf{H} - (q\hbar/2Mc)(\mathbf{p}\cdot\mathbf{A} + \mathbf{A}\cdot\mathbf{p}) + (q^2/2Mc^2)\mathbf{A}^2, \qquad (8.1)$$

where $\mathbf{A}$ and ϕ stand for the magnetic and electric potentials, and where the magnetic field $\mathbf{H}$ satisfies the usual relation $\mathbf{H} = \nabla \times \mathbf{A}$. The last two terms in (8.1) come from making the replacement

$$\hbar\mathbf{p} \rightarrow \hbar\mathbf{p} - q\mathbf{A}/c \qquad (8.2)$$

in the kinetic-energy term $\hbar^2\mathbf{p}^2/2M$. For electron t we take

$$(\mathbf{p}qM\boldsymbol{\mu}) \equiv (\mathbf{p}_t, -e, m, -2\beta\mathbf{s}_t), \qquad (8.3)$$

while for nucleus X,

$$(\mathbf{p}qM\boldsymbol{\mu}) \equiv (\mathbf{p}_X, Z_X e, M_X, \beta_N g_X \mathbf{I}_X), \tag{8.4}$$

where $g_X = \mu_X/I_X$. The total contribution to the Hamiltonian of the molecule is obtained by making the replacements (8.3) and (8.4) in (8.1), and then summing over t (which runs from 1 to n) and the two values A and B of X.

Three cases are of particular interest to us:

(i) The molecule is subjected to a constant electric field $\mathbf{E}$.
(ii) The molecule is subjected to a constant magnetic field $\mathbf{H}$.
(iii) The molecule interacts with a plane electromagnetic wave characterized by a propagation vector $\mathbf{k}$ and a polarization vector $\boldsymbol{\pi}$.

Each is now taken up in turn.

8.2 THE STARK EFFECT

Since $\mathbf{E} = -\nabla\phi$, a constant electric field can be obtained by setting $\phi = -\mathbf{E}\cdot\mathbf{r}$. The expression (8.1), when summed over the particles, yields the Stark Hamiltonian H_S:

$$\begin{aligned} H_S &= \sum_t e\mathbf{E}\cdot\mathbf{r}_t - Z_A e\mathbf{E}\cdot\mathbf{r}_A - Z_B e\mathbf{E}\cdot\mathbf{r}_B \\ &= -\mathbf{E}\cdot\mathbf{d}, \end{aligned} \tag{8.5}$$

where $\mathbf{d}$, the electric dipole moment of the molecule, is given by

$$\mathbf{d} = -e\sum_t \mathbf{r}_t + eZ_A\mathbf{r}_A + eZ_B\mathbf{r}_B. \tag{8.6}$$

If the net charge $Z_T e$ of the molecule is zero, i.e., if

$$Z_T = Z_A + Z_B - n = 0, \tag{8.7}$$

then $\mathbf{d}$ is independent of the origin of coordinates. All cases of interest to us satisfy $Z_T = 0$, and a convenient choice for the origin is C, the center of mass of the nuclear frame. Under the successive transformations (6.4) and (6.8), the dipole moment becomes

$$\mathbf{d} = -e\sum_t \boldsymbol{\varrho}_t - e\left(\frac{Z_A M_B - Z_B M_A}{M_A + M_B}\right) r\mathbf{C}^{(1)}(\theta\phi). \tag{8.8}$$

As can be seen immediately from Eq. (8.6), the dipole moment $\mathbf{d}$ is odd under inversion. Since the molecular states that we have been using are characterized by well-defined parity, all diagonal matrix elements of H_S

must vanish. We might be tempted to conclude that no energy-level displacements linear in **E** can occur. However, the analysis of Section 6.11 led to degenerate pairs of kets of opposite parity; and although residual perturbations may produce small splittings (see Section 9.2), a linear Stark effect can occur for sufficiently large E provided H_S can connect the two states of opposing parity belonging to a near-degenerate pair.

Since **d** is a spin-independent vector, its selection rules run

$$\Delta\Sigma = 0, \qquad \Delta\Lambda = 0, \pm 1, \qquad \Delta\Omega = 0, \pm 1$$

for Hund's case (a), and

$$\Delta M_S = 0, \qquad \Delta\Lambda = 0, \pm 1$$

for Hund's case (b). When **d** is set between states of the type (6.63) or (6.64), the only nonzero contributions come from bra and ket possessing identical values of Λ. For Hund's case (a), we find

$$\langle a+ \mid H_S \mid a- \rangle = -\langle [\nu\Lambda S\Sigma][v][JM_J\ -\Omega] \mid \mathbf{E}\cdot\mathbf{d} \mid [\nu\Lambda S\Sigma][v][JM_J\ -\Omega] \rangle. \tag{8.9}$$

It is convenient to take **E** along the z direction of the laboratory frame F, for then

$$E_z d_z = E d_{00}{}^{(10)} = E(\mathbf{D}^{(11)}\mathbf{d}^{(01)})_{00}{}^{(10)}$$
$$\equiv E D_{00}{}^{(11)} d_{00}{}^{(01)} (10, 10 \mid 00),$$

since, for the problem in hand, only the component $D_{00}{}^{(11)}$ of $\mathbf{D}^{(11)}$ can contribute to the matrix element of Eq. (8.9). Applying the WE theorem, we get

$$\langle a+ \mid H_S \mid a- \rangle = E(3)^{-1/2}(-1)^{M_J+\Omega} \begin{pmatrix} J & 1 & J \\ -M_J & 0 & M_J \end{pmatrix} \begin{pmatrix} J & 1 & J \\ \Omega & 0 & -\Omega \end{pmatrix} (J \,\|\, D^{(11)} \,\|\, J)\mu_e, \tag{8.10}$$

where

$$\mu_e = \langle [\nu\Lambda][v] \mid d_{00}{}^{(01)} \mid [\nu\Lambda][v] \rangle. \tag{8.11}$$

From Eq. (8.8), we may readily show that

$$d_{00}{}^{(01)} = -e \sum_t \zeta_t - eZ_A r_{AC} + eZ_B r_{BC},$$

which makes it obvious that μ_e is the electric dipole moment of the molecule in the frame F' for either state $\mid [\nu \pm\Lambda][v] \rangle$. The 3-$j$ symbols and the reduced matrix element in Eq. (8.10) can be immediately evaluated to yield

$$\langle a+ \mid H_S \mid a- \rangle = -E\mu_e M_J \Omega / J(J+1). \tag{8.12}$$

In the presence of a small splitting 2δ between companion states of opposite parity, the secular equation for the energies W is

$$\begin{vmatrix} \delta - W & \epsilon \\ \epsilon & -\delta - W \end{vmatrix} = 0,$$

where $\epsilon = \langle a+ \mid H_S \mid a- \rangle$. For $\delta = 0$ (or, alternatively, in the limit of large electric fields), the energies W are given by $\pm\epsilon$. In fact, $+\epsilon$ is the energy of the component whose eigenket is

$$| \, [\nu\Lambda S\Sigma][v][JM_J \, -\Omega]\rangle;$$

and $-\epsilon$ corresponds to

$$| \, [\nu \, -\Lambda \; S \; -\Sigma][v][JM_J\Omega]\rangle.$$

Except when $\Lambda = \Sigma = \Omega = M_J = 0$, these components are doubly degenerate, since their energies (as given in Eq. (8.12)) are invariant under the simultaneous replacements $\Omega \to -\Omega$, $M_J \to -M_J$. We have seen (in Problem 6.5) that this pair of sign changes corresponds to time reversal; and it is obvious on general grounds that if the total Hamiltonian is invariant under time reversal—as indeed it is in the present case—then every eigenfunction ψ and its time-reversed companion $T\psi$ must correspond to the same energy. Provided ψ and $T\psi$ are distinct, this double degeneracy is preserved when higher order interactions are taken into account. When the molecule possesses an odd number of electrons, J is necessarily half-integral and there is no possibility that $T\psi$ is a multiple of ψ. The double degeneracy is now referred to as Kramers' degeneracy (see Heine [101]).

A good example of the molecular Stark effect is provided by the work of Klemperer and his associates [121] on the microwave spectrum of the lowest $^3\Pi$ state of carbon monoxide. The coupling approximates reasonably well to Hund's case (a), and results have been reported for a number of vibrational and rotational states. By enlarging the Hamiltonian to allow for a variety of perturbations (especially those coming from the adjacent $^3\Sigma^+$ state), a virtually perfect agreement can be obtained between theory and experiment by adjusting a limited set of parameters. Accurate values of the electric dipole moment μ_e for the components of $^3\Pi$ for which $v = 0$, 1, 2, and 3 have been found, as well as the variation of μ_e with J.

8.3 THE ZEEMAN EFFECT

A constant magnetic field **H** is conveniently obtained from a vector potential **A** given by $\frac{1}{2}\mathbf{H} \times \mathbf{r}$. In the first instance, our attention is directed to terms in the Hamiltonian (8.1) that are linear in **H**; we therefore set aside the term involving $\mathbf{A}^2$ for the time being. Since $\mathbf{A}$ satisfies $\nabla\cdot\mathbf{A} = 0$ (a condition often referred to as the Coulomb gauge), we can write $\mathbf{p}\cdot\mathbf{A} = \mathbf{A}\cdot\mathbf{p}$. Thus

$$\mathbf{p}\cdot\mathbf{A} + \mathbf{A}\cdot\mathbf{p} = (\mathbf{H}\times\mathbf{r})\cdot\mathbf{p} = \mathbf{H}\cdot(\mathbf{r}\times\mathbf{p}) = -i\mathbf{H}\cdot(\mathbf{r}\times\nabla).$$

Equations (6.5) can be used to provide the appropriate forms of ∇ for the electrons and the nuclei. The linear combination

$$(Z_A/M_A)(\mathbf{r}_A\times\nabla_A) + (Z_B/M_B)(\mathbf{r}_B\times\nabla_B) - \sum_t (1/m)(\mathbf{r}_t\times\nabla_t), \tag{8.13}$$

when converted to new coordinates according to the scheme (6.4), yields many cross terms involving such combinations as $\mathbf{r}_{AB}\times\nabla_{tC}$, $\mathbf{r}_{tC}\times\nabla_O$, etc. Each one of these vector products comprises two vectors that refer to different coordinates and that possess odd parity. All their diagonal matrix elements must therefore vanish. Setting these terms aside, we find that the expression (8.13) reduces to

$$\frac{Z_A + Z_B - n}{M_A + M_B + mn}(\mathbf{r}_O\times\nabla_O) + \frac{Z_B M_A^2 + Z_A M_B^2}{M_A M_B(M_A + M_B)}(\mathbf{r}_{BA}\times\nabla_{BA})$$
$$-\frac{1}{m}\sum_t(\mathbf{r}_{tC}\times\nabla_{tC}). \tag{8.14}$$

The first term describes the spiraling in the magnetic field of the molecule as a whole when the net electric charge eZ_T does not vanish. This is of no interest to us and is dropped. The remaining terms are directly related to the angular momentum $\mathbf{l}_{BA}$ $(=\mathbf{R})$ of the nuclear motion and $\mathbf{L}$, the total angular momentum of the electrons. The prefacing coefficient m^{-1} of the last term in (8.14) should, strictly speaking, be

$$m^{-1} - (M_A + M_B)^{-1}$$

when $Z_T = 0$. There is no point in making this minor correction in the face of our neglect of relativistic effects. (It is interesting to notice that for positronium, for which $M_B = 0$ and $M_A = m$, this term vanishes—as does the entire orbital contribution to the Zeeman Hamiltonian.) We can now write down the Zeeman Hamiltonian H_Z. Including the interaction between $\mathbf{H}$ and the magnetic moments of the particles, we have

$$H_Z = -g_r\beta_N\mathbf{H}\cdot\mathbf{R} + \beta\mathbf{H}\cdot(\mathbf{L} + 2\mathbf{S}) - g_A\beta_N\mathbf{H}\cdot\mathbf{I}_A - g_B\beta_N\mathbf{H}\cdot\mathbf{I}_B, \tag{8.15}$$

where

$$g_r = M_P(Z_A M_B^2 + Z_B M_A^2)/M_A M_B(M_A + M_B). \tag{8.16}$$

For Hund's cases (a) and (b), we replace **R** by **J** − **P** and **N** − **L** respectively.

The principal term on the right-hand side of Eq. (8.15) is the second one, representing the interaction between the electrons and the magnetic field. Since **L** + 2**S** is a vector, the analysis for Hund's case (a) proceeds in a similar manner to that for the Stark effect: the main difference is that H_Z conserves parity. Thus Eq. (8.9) is replaced by

$$\begin{aligned}\langle a+ \mid H_Z \mid a+\rangle \\ &= \langle a- \mid H_Z \mid a-\rangle \\ &= \beta\langle[\nu\Lambda S\Sigma][v][JM_J\,-\Omega] \mid \mathbf{H}\cdot(\mathbf{L}+2\mathbf{S}) \mid [\nu\Lambda S\Sigma][v][JM_J\,-\Omega]\rangle,\end{aligned}$$

and from this point the analysis proceeds as before. We have merely to make the replacement

$$\begin{aligned}-E\mu_e &\rightarrow H\beta\langle[\nu\Lambda S\Sigma][v] \mid L_\zeta + 2S_\zeta \mid [\nu\Lambda S\Sigma][v]\rangle \\ &= H\beta(\Lambda + 2\Sigma).\end{aligned}$$

Since the secular determinant for H_Z is diagonal, the effect of a splitting 2δ between companion levels of opposite parity is to add and subtract δ to and from the common value of $\langle a\pm \mid H_Z \mid a\pm\rangle$. The energies of the components of a level relative to the average energy of the pair of levels in zero field are

$$\frac{H\beta(\Lambda + 2\Sigma)M_J\Omega}{J(J+1)} \pm \delta. \tag{8.17}$$

The zero-field splitting is thus superposed on every component of the Zeeman multiplet and all the degeneracy is thereby removed. The possibility of lifting the degeneracy stems from the fact that H_Z is of the form **H**·**Z**, where the vector **Z**, unlike its electric analog **d**, changes sign under time reversal. The invariance of the Hamiltonian with respect to time reversal is lost, and we can no longer draw any conclusions as to the degeneracy of the levels.

A consequence of the expression (8.17) is that $^2\Pi_{1/2}$ states (for which the subscript $\frac{1}{2}$ denotes the value of $|\Omega|$) are nonmagnetic, since $\Lambda = \pm 1$, $\Sigma = \mp\frac{1}{2}$. However, deviations from pure case (a) coupling are common. For example, the Zeeman splittings that Radford [122] has observed for the $^2\Pi_{1/2}$ state of OH are largely due to this effect.

An interesting situation arises for $^1\Sigma$ states. Since $\Lambda = \Sigma = 0$, the electronic part of the Zeeman Hamiltonian makes no diagonal contribution, and the smaller terms in Eq. (8.15) are exposed. The first one, namely $-g_r\beta_N\mathbf{H}\cdot\mathbf{R}$, is of particular interest, since an explicit form has been given for g_r. For the molecules H_2, HD, and D_2, which possess identical electronic structures, the theory gives

$$g_r(H_2) = 1, \qquad g_r(HD) = \tfrac{5}{6}, \qquad g_r(D_2) = \tfrac{1}{2}. \tag{8.18}$$

The precise experimental results of Ramsey [119] are

$$g_r(H_2) = 0.8829, \qquad g_r(HD) = 0.6601, \qquad g_r(D_2) = 0.4406. \tag{8.19}$$

The agreement is only moderate. What is remarkable, however, is that the experimental figures are proportional to the inverses of the molecular reduced masses, i.e., to $(M_A + M_B)/M_A M_B$, rather than to the theoretical expression for g_r given in (8.14). Evidently second order effects involving excited electronic states are important, and they must somehow conspire to change the proportionality. This point is taken up again in Section 9.3.

8.4 DIAGONAL CORRECTIONS TO H_Z

Additional contributions to the Zeeman Hamiltonian H_Z can occur as soon as deviations from the nonrelativistic limit are admitted. The Breit interaction is the principal source of such terms. We have merely to make replacements of the type (8.2) and write $\frac{1}{2}(\mathbf{H} \times \mathbf{r})$ for $\mathbf{A}$ to obtain contributions to H_Z that are smaller than the leading term $\beta\mathbf{H}\cdot(\mathbf{L} + 2\mathbf{S})$ by roughly $\alpha^2 \mid q_1 q_2 \mid$, where $q_1 e$ and $q_2 e$ are the charges of a pair of interacting particles, and α is the fine-structure constant. An account of these additional terms has been given by Carrington *et al.* [35]. We shall restrict ourselves to an illustration of the general method by taking the contribution that comes from the second part of H_{so} given in Eq. (7.1), namely

$$-2\beta^2 \sum_{tX} [\nabla_t (Z_X r_{tX}^{-1}) \times \mathbf{p}_t]\cdot \mathbf{s}_t. \tag{8.20}$$

To get the additional contribution to H_Z, we make the replacement

$$\mathbf{p}_t \rightarrow \mathbf{p}_t + (e/2\hbar c)(\mathbf{H} \times \mathbf{r}_t). \tag{8.21}$$

The expression (8.20) becomes augmented by H_Z', where

$$H_Z' = \tfrac{1}{2}\beta\alpha^2 a_0 \sum_{tX} Z_X r_{tX}^{-3} \{\mathbf{r}_{tX} \times (\mathbf{H} \times \mathbf{r}_t)\}\cdot \mathbf{s}_t. \tag{8.22}$$

Elementary vector analysis allows us to write

$$\{\mathbf{r}_{tX} \times (\mathbf{H} \times \mathbf{r}_t)\}\cdot\mathbf{s}_t = \mathbf{H}\cdot\{\mathbf{s}_t(\mathbf{r}_{tX}\cdot\mathbf{r}_t) - \mathbf{r}_{tX}(\mathbf{r}_t\cdot\mathbf{s}_t)\}. \tag{8.23}$$

If we now assume that the center of gravity of the nuclear frame is at rest, we can conveniently take it as the origin of coordinates. Thus $\mathbf{r}_t$ is replaced by $\mathbf{r}_{tC}$, and all radial quantities in H_Z' involve relative coordinates in the molecular frame F'. This makes the analysis reasonably straightforward. Things are particularly simple for Hund's case (a) (see Problem 8.2). For Hund's case (b), it is useful to exhibit the tensorial character of the second term in (8.23) by making the recoupling

$$\mathbf{r}_{tX}(\mathbf{r}_{tC}\cdot\mathbf{s}_t) = \sum_k (-1)^{k+1}\{(2k+1)/3\}^{1/2}\{(\mathbf{r}_{tX}\mathbf{r}_{tC})^{(k)}\mathbf{s}_t^{(1)}\}^{(1)}.$$

This puts H_Z' in a convenient form when evaluating it between coupled states of the type $|(NS)JM_J\rangle$. The only complication lies in the sum over t, which runs over both spin and orbital quantities. When dealing with the $^3\Pi$ state of H_2, we replaced $\mathbf{s}_t$ by $\mathbf{S}/2$. Analogous simplifications are available whenever our interest lies solely in matrix elements diagonal with respect to S and, at the same time, S possesses its maximum value for the electrons in unpaired orbitals. In these cases, the electronic eigenfunction can be factored into a spin part and an orbital part; but in general this separation is not possible.

Suppose we are dealing with a situation where we can write $\mathbf{s}_t = \kappa\mathbf{S}$, where κ is some constant independent of t. In this case,

$$H_Z' = \tfrac{1}{2}\beta\alpha^2\kappa a_0 \sum_{tX} \mathbf{H}\cdot[\mathbf{S}(\mathbf{r}_{tX}\cdot\mathbf{r}_{tC}) - \mathbf{r}_{tX}(\mathbf{r}_{tC}\cdot\mathbf{S})]r_{tX}^{-3}. \qquad (8.24)$$

As already mentioned, the coordinates of electron t appear here relative to points set in the molecular frame F'. This means that when H_Z' is set between a bra and a ket, integration can be immediately carried out over the coordinates $\boldsymbol{\rho}_t$ and r. By doing this—at least in principle—we can see more clearly how H_Z' can be parametrized. We begin with a recoupling:

$$H_Z' = \tfrac{1}{2}\beta\alpha^2\kappa a_0 \sum_{tX} [\tfrac{2}{3}(\mathbf{H}\cdot\mathbf{S})(\mathbf{r}_{tX}\cdot\mathbf{r}_{tC}) - (\mathbf{HS})^{(2)}\cdot(\mathbf{r}_{tX}\mathbf{r}_{tC})^{(2)}]r_{tX}^{-3}. \qquad (8.25)$$

Terms involving $(\mathbf{r}_{tX}\mathbf{r}_{tC})^{(1)}$ have been dropped, since the ζ component vanishes, and the ξ and η components behave as shift operators for which $\Delta\Lambda = \pm 1$. These can produce no effect within a given electronic term, since Λ and $-\Lambda$ must necessarily differ by a multiple of 2. For a similar reason, we do not require the ± 1 components of the tensors of rank 2 in Eq. (8.25). Expanding the remaining parts, we get

$$H_Z' = \Delta g_\xi \beta H_\xi S_\xi + \Delta g_\eta \beta H_\eta S_\eta + \Delta g_\zeta \beta H_\zeta S_\zeta, \qquad (8.26)$$

where

$$\Delta g_\zeta = \tfrac{1}{2}\alpha^2\kappa a_0 \langle[\nu\Lambda]| \sum_{tX} (\xi_t^2 + \eta_t^2) r_{tX}^{-3} |[\nu\Lambda]\rangle,$$

$$\tfrac{1}{2}(\Delta g_\xi + \Delta g_\eta) = \tfrac{1}{2}\alpha^2\kappa a_0 \langle[\nu\Lambda]| \sum_{tX} (\zeta_{tX}\zeta_{tC} + \tfrac{1}{2}\xi_t^2 + \tfrac{1}{2}\eta_t^2) r_{tX}^{-3} |[\nu\Lambda]\rangle,$$

$$\tfrac{1}{2}(\Delta g_\xi - \Delta g_\eta) = -\tfrac{1}{2}\alpha^2\kappa a_0 \langle[\nu\Lambda]| \sum_{tX} (\xi_t^2 - \eta_t^2) r_{tX}^{-3} |[\nu -\Lambda]\rangle. \qquad (8.27)$$

This shows that only three essentially distinct radial integrals appear in the analysis. We preface g_ξ, g_η, and g_ζ with Δ to emphasize that we are studying corrections to the leading part H_Z of the total Zeeman Hamiltonian. The last matrix element in the set (8.27) is nonzero only for Π states: for all the others we can write

$$\Delta g_\zeta = \Delta g_{||}, \qquad \Delta g_\xi = \Delta g_\eta = \Delta g_\perp, \tag{8.28}$$

so that now Eq. (8.26) becomes

$$H_Z' = \Delta g_{||}\beta H_\zeta S_\zeta + \Delta g_\perp \beta(H_\xi S_\xi + H_\eta S_\eta). \tag{8.29}$$

A convenient way to handle this last equation is to put it in the form

$$H_Z' = \Delta g_\perp \beta \mathbf{H}\cdot\mathbf{S} + (\Delta g_{||} - \Delta g_\perp)\beta H_\zeta S_\zeta$$

and then transform H_ζ and S_ζ into the frame F of the laboratory. If $\mathbf{H}$ is taken parallel to the z axis, we have

$$H_\zeta = H_{00}{}^{(01)} = (\mathbf{D}^{(11)}\mathbf{H}^{(10)})_{00}{}^{(01)} = -(3)^{-1/2}HD_{00}{}^{(11)}.$$

Thus

$$\begin{aligned} H_\zeta S_\zeta &= -(3)^{-1/2}H \sum_k (10, 10 \mid k0)\{\mathbf{D}^{(11)}(\mathbf{D}^{(11)}\mathbf{S}^{(10)})^{(01)}\}_{00}{}^{(1k)} \\ &= \tfrac{1}{3}HS_z - \tfrac{1}{3}(2)^{1/2}H(\mathbf{D}^{(22)}\mathbf{S}^{(10)})_{00}{}^{(12)}. \end{aligned}$$

Putting the two parts together, we have, for non-Π states,

$$H_Z' = \tfrac{1}{3}(\Delta g_{||} + 2\Delta g_\perp)\beta HS_z + \tfrac{1}{3}(2)^{1/2}(\Delta g_\perp - \Delta g_{||})\beta H(\mathbf{D}^{(22)}\mathbf{S}^{(10)})_{00}{}^{(12)}. \tag{8.30}$$

The matrix elements of the first term are easy to evaluate. As a variant to the usual method of treating the second term, we introduce a null angular momentum through the replacement

$$|[N - \Lambda]SJM_J\rangle \to |[NSJM_J][N0N - \Lambda]\rangle$$

in order to put the spaces F and F' on a similar footing. Then a double application of the WE theorem, together with Eqs. (1.62) and (3.12), leads to the result

$$\begin{aligned} &\langle[N - \Lambda]SJM_J \mid (\mathbf{D}^{(22)}\mathbf{S}^{(10)})^{(12)} \mid [N' - \Lambda]SJ'M_J'\rangle \\ &\quad = (-1)^{J-M_J}\begin{pmatrix} J & 1 & J' \\ -M_J & 0 & M_J' \end{pmatrix}(-1)^{N+\Lambda}\begin{pmatrix} N & 2 & N' \\ \Lambda & 0 & -\Lambda' \end{pmatrix}\Phi, \end{aligned}$$

where

$$\begin{aligned}\Phi &= ([NSJ][N0N] \,||\, (\mathbf{D}^{(22)}\mathbf{S}^{(10)})^{(12)} \,||\, [N'SJ'][N'0N']) \\ &= \{15[N, N', J, J']\}^{1/2}(N \,||\, D^{(22)} \,||\, N')(S \,||\, S \,||\, S) \\ &\quad \times \begin{Bmatrix} N & N' & 2 \\ 0 & 0 & 0 \\ N & N' & 2 \end{Bmatrix} \begin{Bmatrix} N & N' & 2 \\ S & S & 1 \\ J & J' & 1 \end{Bmatrix} \\ &= \{15S(S+1)[N, S, J, N', J']\}^{1/2} \begin{Bmatrix} N & N' & 2 \\ S & S & 1 \\ J & J' & 1 \end{Bmatrix} \end{aligned} \tag{8.31}$$

Of course, we could get this result without introducing a null angular momentum; but the above method is scarcely more elaborate, and it illustrates how a matrix element can be immediately reduced in both spaces F and F' simultaneously if we so wish.

It is interesting to note that for an atomic electron moving in a central potential Ze/r, the contribution H_Z' to the Zeeman Hamiltonian collapses to

$$\tfrac{1}{2}\beta\alpha^2 a_0 Z\mathbf{H}\cdot[\mathbf{s} - \mathbf{r}(\mathbf{r}\cdot\mathbf{s})r^{-2}]\langle r^{-1}\rangle. \tag{8.32}$$

From the virial theorem, the average kinetic energy T of the electron is the negative of one half the average potential energy. In other words,

$$T = \tfrac{1}{2}\langle Ze^2/r\rangle,$$

and the expression (8.32) becomes

$$\beta\mathbf{H}\cdot[\mathbf{s} - \mathbf{r}(\mathbf{r}\cdot\mathbf{s})r^{-2}]T/mc^2.$$

This is identical to part of what Abragam and Van Vleck [123] call the *Yale* correction for the atomic Zeeman Hamiltonian.

8.5 TRANSITIONS BETWEEN MOLECULAR STATES

The interaction between a diatomic molecule and a radiation field is properly treated by applying quantum theory to the total system of particles plus field. It would be too great a digression to run over the general theory of this subject, which, in any case, is dealt with in texts on quantum mechanics. The aspects of the problem of particular interest to us are the classification of the radiation and the application of the theory to diatomic molecules. The interaction Hamiltonian is, of course, central to the discussion; and the theory takes a particularly simple form if the field is represented by a plane wave characterized by a given polarization vector $\boldsymbol{\pi}$ and propagation vector $\mathbf{k}$. This is an entirely appropriate starting point for the treatment of induced transitions, since the field is at our disposal. The alterations that an analysis of the spontaneous transitions entails lie not in the molecular matrix elements (i.e., those depending on the coordinates of the particles), but solely in the treatment of the field. In other words, all the relevant quantities for us occur in an analysis of the induced transitions, and they reappear in the same form when spontaneous transitions are considered.

A convenient expression for the vector potential $\mathbf{A}$ is the complex form defined by

$$\mathbf{A} = A\boldsymbol{\pi}\exp(i\mathbf{k}\cdot\mathbf{r} - i\omega t), \tag{8.33}$$

where the angular frequency ω is related to the magnitude k of $\mathbf{k}$ through the relation $\omega = ck$. We introduce the space-independent part $\mathbf{A}_0$ of $\mathbf{A}$ by writing

$$\mathbf{A} = \mathbf{A}_0 \exp(i\mathbf{k}\cdot\mathbf{r}),$$

and this notation can be immediately extended to other field vectors. The usual gauge in which $\phi = 0$ is adopted, and the electric vector $\mathbf{E}$ is derived from $\mathbf{A}$ through the relations

$$\mathbf{E} = -\dot{\mathbf{A}}/c = i(\omega/c)\mathbf{A} = ik\mathbf{A}. \tag{8.34}$$

Since $\mathbf{E}$ is perpendicular to $\mathbf{k}$, we can deduce

$$\boldsymbol{\nabla}\cdot\mathbf{A} = i\mathbf{k}\cdot\mathbf{A} = 0,$$

provided, of course, the context makes it clear that the operator $\boldsymbol{\nabla}$ acts solely on $\mathbf{A}$ and not as well on other functions of $\mathbf{r}$ that may stand to the right of $\mathbf{A}$. If we fix our attention on a particle of charge q, mass M, and magnetic moment $\boldsymbol{\mu}$, the Hamiltonian h_{int} describing the interaction of this particle with the field is given by

$$h_{\text{int}} = -(q\hbar/Mc)\mathbf{A}\cdot\mathbf{p} - \boldsymbol{\mu}\cdot\mathbf{H}, \tag{8.35}$$

where $\mathbf{A}$ is specified by Eq. (8.33). The general expansion of $\exp(i\mathbf{k}\cdot\mathbf{r})$ can be effected by means of Eq. (5.6). For the applications that we have in mind, however, the wavelength λ (given by $2\pi/k$) is very much larger than the dimensions of the molecule. Provided, then, that the origin for $\mathbf{r}$ is chosen in the general vicinity of the molecule, we have $\mathbf{k}\cdot\mathbf{r} \ll 1$. It is now an adequate approximation to make the replacement

$$\exp(i\mathbf{k}\cdot\mathbf{r}) \to 1 + i(\mathbf{k}\cdot\mathbf{r}). \tag{8.36}$$

A stratagem allows $\mathbf{p}$ (in Eq. (8.35)) to be replaced by a more amenable operator. We first note that the eventual use of h_{int}, when summed over the particles (thereby giving H_{int}), lies in setting it between two eigenstates $\langle a \mid$ and $\mid b\rangle$ of the molecular Hamiltonian H. (Throughout this section, the symbols a and b do not refer to the Hund coupling cases.) We denote the corresponding eigenvalues by E_a and E_b respectively. In the nonrelativistic limit, the only part of H that does not commute with the radius vector $\mathbf{r}$ of a particle is the kinetic-energy term $\hbar^2\mathbf{p}^2/2M$ of the same particle. So

$$\begin{aligned}(E_a - E_b)\langle a \mid \mathbf{r} \mid b\rangle &= \langle a \mid [H, \mathbf{r}] \mid b\rangle \\ &= \langle a \mid [\hbar^2\mathbf{p}^2/2M, \mathbf{r}] \mid b\rangle = \langle a \mid (\hbar^2\mathbf{p}/iM) \mid b\rangle.\end{aligned} \tag{8.37}$$

The time dependences associated with $\langle a \mid$, $\mathbf{A}$, and $\mid b\rangle$ are

$$\exp(iE_a t/\hbar), \qquad \exp(-i\omega t), \qquad \exp(-iE_b t/\hbar)$$

respectively, from which we obtain the Bohr frequency condition

$$E_a - E_b = \hbar\omega = \hbar ck \tag{8.38}$$

without any ambiguity in sign. Combining Eqs. (8.37) and (8.38), we arrive at the off-diagonal operator equivalence

$$\hbar\mathbf{p} \equiv ickM\mathbf{r}. \tag{8.39}$$

A purely classical analog can be easily derived by equating the momentum to $M\dot{r}$ and assuming $r = r_0 e^{i\omega t}$. It should also be mentioned that Eqs. (8.37) amount to what is sometimes called a hypervirial theorem [124].

For the first term in the expansion (8.36), we now have

$$-(q\hbar/Mc)\mathbf{A}\cdot\mathbf{p} = -ikqA\boldsymbol{\pi}\cdot\mathbf{r}e^{-i\omega t} = -q\mathbf{E}_0\cdot\mathbf{r} \tag{8.40}$$

from Eq. (8.34). Summing over all the particles that make up the molecule, we get the Hamiltonian H_{ed} for electric-dipole radiation:

$$H_{\text{ed}} = -\mathbf{E}_0\cdot\mathbf{d}, \tag{8.41}$$

where $\mathbf{d}$ is defined, as before, through Eq. (8.6).

The second term in the replacement (8.36) yields

$$-(q\hbar/Mc)Ai(\mathbf{k}\cdot\mathbf{r})(\boldsymbol{\pi}\cdot\mathbf{p})e^{-i\omega t} = -(q\hbar/Mc)i(\mathbf{k}\cdot\mathbf{r})(\mathbf{A}_0\cdot\mathbf{p}). \tag{8.42}$$

A recoupling is called for to bring together the operators that act on the coordinates of the particle. We use

$$(\mathbf{k}\cdot\mathbf{r})(\mathbf{A}_0\cdot\mathbf{p}) = \sum_x (-1)^x(\mathbf{kA}_0)^{(x)}\cdot(\mathbf{rp})^{(x)}. \tag{8.43}$$

Since $\mathbf{k}$ is perpendicular to $\mathbf{A}_0$, the term for which $x = 0$ vanishes. For $x = 1$,

$$-(\mathbf{kA}_0)^{(1)}\cdot(\mathbf{rp})^{(1)} = \tfrac{1}{2}(\mathbf{k}\times\mathbf{A}_0)\cdot(\mathbf{r}\times\mathbf{p}).$$

Now

$$\mathbf{H} = \nabla\times\mathbf{A} = i\mathbf{k}\times\mathbf{A};$$

so the term for which $x = 1$ contributes

$$-(q\hbar/2Mc)\mathbf{H}_0\cdot\boldsymbol{l}$$

to the right-hand side of Eq. (8.42). The second term $-\boldsymbol{\mu}\cdot\mathbf{H}$ of h_{int} can be replaced by $-\boldsymbol{\mu}\cdot\mathbf{H}_0$ to zeroth order in powers of $(\mathbf{k}\cdot\mathbf{r})$, and we get, in all

$$-\mathbf{H}_0\cdot[(q\hbar/2Mc)\boldsymbol{l} + \boldsymbol{\mu}]. \tag{8.44}$$

When this is summed over all the particles comprising the molecule, the Hamiltonian H_{md} for magnetic-dipole radiation is obtained. To the order at which we are working, we may use $\mathbf{H}$ in place of $\mathbf{H}_0$, and this substitution makes H_{md} formally identical to the Hamiltonian from which H_{Z} was derived (in Section 8.3), though now, of course, $\mathbf{H}$ is time-dependent. Before concluding that we can write $H_{\text{md}} = H_{\text{Z}}$, where H_{Z} is defined as in Eq. (8.15), we should recall that the parity arguments that led us to discard certain cross terms in the derivation of Eq. (8.15) no longer hold good, since our interest in H_{md} necessarily lies in its off-diagonal matrix elements rather than its diagonal ones. However, the cross terms are of the same order of magnitude as those in Eq. (8.15) that involve β_N rather than β. Setting all of these terms aside, we have

$$H_{\text{md}} = \beta\mathbf{H}_0\cdot(\mathbf{L} + 2\mathbf{S}). \tag{8.45}$$

Of course, if the matrix elements of $\mathbf{L} + 2\mathbf{S}$ vanish, then the neglected terms may become significant (as they do in nuclear magnetic resonance). The close similarity between H_{md} and H_{Z}, as well as that between H_{ed} and H_{S}, can be readily understood by noting that the electromagnetic wave goes over into the static-field case in the limit $\mathbf{k}\to 0$.

There remains the term corresponding to $x = 2$, which contributes

$$-(q\hbar/Mc)i(\mathbf{kA}_0)^{(2)}\cdot(\mathbf{rp})^{(2)} \tag{8.46}$$

to h_{int}. By analogous methods that led to the operator equivalence (8.39), we obtain

$$\hbar(\mathbf{rp})^{(2)} \equiv ikMc(6)^{-1/2}r^2\mathbf{C}^{(2)}. \tag{8.47}$$

(The general equivalence is given in Problem 8.4.) The result of summing (8.46) over the particles that make up the molecule defines H_{eq}, the Hamiltonian for electric quadrupole radiation. Using the equivalence (8.47), we get

$$H_{\text{eq}} = k(6)^{-1/2}(\mathbf{k}\mathbf{A}_0)^{(2)}\cdot\mathbf{Q}^{(2)}, \tag{8.48}$$

where the molecular quadrupole tensor is given by

$$\mathbf{Q}^{(2)} = -e\sum_t r_t^2\mathbf{C}^{(2)}(\theta_t\phi_t) + e\sum_X Z_X r_X^2\mathbf{C}^{(2)}(\theta_X\phi_X). \tag{8.49}$$

To the order at which the expansion (8.36) is valid, we have, for the total interaction Hamiltonian,

$$H_{\text{int}} = H_{\text{ed}} + H_{\text{md}} + H_{\text{eq}}. \tag{8.50}$$

We can now make a direct appeal to time-dependent perturbation theory to obtain the probability w per unit time for an induced transition between two (time-independent) states $|\,a\rangle$ and $|\,b\rangle$ of the molecule. The result is [125]

$$w = (2\pi/\hbar)\;|\,\langle a\,|\,H_{\text{int}}\,|\,b\rangle\,|^2\,\delta(\hbar\omega - E_a + E_b), \tag{8.51}$$

where the Dirac delta function on the right finds its ultimate use in an integration over the frequency distribution of the radiation field. Induced emission or absorption corresponds to taking $|\,a\rangle$ as the initial or final state respectively. From H_{int}, we abstract those parts that depend solely on $\mathbf{k}/k$, $\boldsymbol{\pi}$, and the coordinates of the particles making up the molecule. They are conveniently defined as

$$\begin{aligned} V_{\text{ed}} &= \boldsymbol{\pi}\cdot\mathbf{d}, \\ V_{\text{md}} &= \beta k^{-1}(\mathbf{k}\times\boldsymbol{\pi})\cdot(\mathbf{L}+2\mathbf{S}), \\ V_{\text{eq}} &= (6)^{-1/2}k^{-1}(\mathbf{k}\boldsymbol{\pi})^{(2)}\cdot\mathbf{Q}^{(2)}. \end{aligned} \tag{8.52}$$

As interference between H_{ed}, H_{md}, and H_{eq} is comparatively rare, each part can be treated separately. For any operator V of the three given in Eqs. (8.52), the quantity $\mathcal{S}(ab)$, defined by

$$\mathcal{S}(ab) = |\,\langle a\,|\,V\,|\,b\rangle\,|^2, \tag{8.53}$$

is called the *strength* of the transition. In this form, $\mathcal{S}(ab)$ depends on $\boldsymbol{\pi}$ and $\mathbf{k}/k$. In itself, $\mathcal{S}(ab)$ cannot determine the intensity of an observed transition, since the latter also depends on the nature of the radiation field and the distribution of the molecules among the energy eigenstates. However, under many conditions of practical importance (for example, a small frequency range) the strengths of the transitions give relative intensities.

It is important to recognize that $\langle a \mid$ and $\mid b\rangle$ are eigenstates of the total molecule and as such include the nuclear eigenstates $\mid (I_A I_B) T M_T\rangle$. These normally play no role in the calculations because V is independent of nuclear spin. However, in the case of homonuclear diatomics, there is a correlation between T and the eigenvalues of the nuclear interchange operator O. The latter depend on quantum numbers such as J and N (see Eqs. (6.65)–(6.68)), so the nuclear degeneracy can alternate as J or N runs sequentially. This leads to the alternating intensities that are a striking feature of the spectra of molecules such as N_2. Although little more than a curiosity now, this alternation was of great value in establishing nuclear spins in the 1920s.

8.6 ELECTRIC-DIPOLE TRANSITIONS

The leading term of the replacement (8.36) gives rise to the electric-dipole operator V_{ed} of Eqs. (8.52). This is responsible for all observed electronic transitions save the very weakest. The study of transition strengths presents us for the first time with the calculation of off-diagonal matrix elements. It is usually not difficult to write down many selection rules merely by inspection. For V_{ed}, we make the preliminary observation that $\boldsymbol{\pi}$ does not enter in the integration over the molecular coordinates; thus the selection rules are determined by the properties of $\mathbf{d}$. Since $\mathbf{d}$ is independent of spin, we can begin at once with

$$\Delta S = \Delta\Sigma = \Delta M_S = 0.$$

Turning to Eq. (8.6), we see that $\mathbf{d}$ changes sign under inversion: it follows that electric-dipole transitions can only take place between states of opposite parity. We can go further if we first note that a change of origin adds to $\mathbf{d}$ a constant term whose off-diagonal matrix elements necessarily vanish. It is highly convenient to take the center of gravity C of the nuclear frame as the origin for $\mathbf{d}$, for then we can use the transformation

$$\mathbf{d}^{(10)} = (\mathbf{D}^{(11)}\mathbf{d}^{(01)})^{(10)}$$

to separate off the dependence of $\mathbf{d}$ on the orientation of the nuclear frame. By applying the WE theorem to $\mathbf{D}^{(11)}$, we immediately get the selection rules

$$\Delta\Xi = 0, \pm 1 \qquad (\Xi \equiv J, \Omega, M_J, N, \Lambda, M) \tag{8.54}$$

together with the restriction that the rules $\Delta J = 0$ and $\Delta N = 0$ no longer hold when $J = 0$ and $N = 0$ respectively. These conditions can be strength-end when both bra and ket of a transition matrix element are labeled by zero values of Ω (for Hund's case (a)) or zero values of Λ (for Hund's case (b)). In these instances, an application of the WE theorem in F' yields the 3-j symbols

$$\begin{pmatrix} J' & 1 & J \\ 0 & 0 & 0 \end{pmatrix} \quad \text{and} \quad \begin{pmatrix} N' & 1 & N \\ 0 & 0 & 0 \end{pmatrix},$$

both of which vanish unless the sums of the arguments in the upper rows are even. We obtain

$$\Delta J = \pm 1 \qquad (\Omega = \Omega' = 0)$$

for Hund's case (a), and

$$\Delta N = \pm 1 \qquad (\Lambda = \Lambda' = 0)$$

for Hund's case (b). Thus the transitions for which $\Delta J = 0$ and $\Delta N = 0$ disappear.

By choosing C as the origin for $\mathbf{d}$, the component $d_{00}^{(01)}$ of $\mathbf{d}^{(01)}$ is invariant under the reflection operation R of Section 6.9. Thus no transitions can take place between states of the type Σ^+ and Σ^-. For molecules possessing two identical nuclei, the fact that $\mathbf{d}^{(01)}$ changes sign under inversion eliminates all transitions except those of the type $g \leftrightarrow u$.

All selection rules emerge from an actual calculation of the transition strengths, of course. To illustrate the general method, we take the transitions corresponding to ${}^1\Sigma \to {}^1\Pi$, which have been studied by Crawford [126] with particular reference to the so-called Ångström bands of CO. It is supposed that a constant magnetic field is applied to break up every J level of ${}^1\Pi$ into $2J + 1$ components. Reference to the expression (8.17) shows that no J' level of ${}^1\Sigma$ is split: hence the transitions give directly the Zeeman splittings of ${}^1\Pi$. The direction of the Zeeman field $\mathbf{H}$ is taken to define the z axis. For a component π_{-q} of $\boldsymbol{\pi}$, we have to calculate the matrix element of d_q when set between the bra

$$\langle [{}^1\Sigma][v'][J'M'0] | \tag{8.55}$$

and one of the two kets

$$\sqrt{\tfrac{1}{2}}\, \{ | [{}^1\Pi\, 1][v][JM\, -1] \rangle \pm | [{}^1\Pi\, -1][v][JM1] \rangle \}. \tag{8.56}$$

Since these kets possess opposite parity, only one linear combination can give a nonzero matrix element; and, for this one, the matrix element is

$$(2)^{1/2} \langle [{}^1\Sigma][v'][J'M'0] | d_q | [{}^1\Pi\, 1][v][JM\, -1] \rangle.$$

TABLE 8.1 Strengths $\mathcal{S}_{ed}$ for Transitions $^1\Sigma \to {}^1\Pi$

Branch	$\parallel$	$\perp$
P	$\mu_{\Sigma\Pi}^2 \dfrac{(J+1)(J^2-M^2)}{J(2J-1)(2J+1)}$	$\mu_{\Sigma\Pi}^2 \dfrac{(J+1)(J^2-J+M^2)}{2J(2J-1)(2J+1)}$
Q	$\mu_{\Sigma\Pi}^2 \dfrac{M^2}{J(J+1)}$	$\mu_{\Sigma\Pi}^2 \dfrac{J^2+J-M^2}{2J(J+1)}$
R	$\mu_{\Sigma\Pi}^2 \dfrac{J[(J+1)^2-M^2]}{(J+1)(2J+1)(2J+3)}$	$\mu_{\Sigma\Pi}^2 \dfrac{J[(J+1)(J+2)+M^2]}{(2J+1)(2J+2)(2J+3)}$

We can now use our standard techniques. For nonvanishing contributions, we write

$$d_q^{(1)} = d_{q0}^{(10)} = (\mathbf{D}^{(11)}\mathbf{d}^{(01)})_{q0}^{(10)}$$

$$\equiv (1\ -1,\ 11 \mid 00)\, D_{q1}^{(11)} d_{0-1}^{(01)},$$

and use the WE theorem to evaluate the matrix elements of $D_{q1}^{(11)}$. The transition strength $\mathcal{S}_{ed}$ corresponding to π_{-q} is

$$2\mu_{\Sigma\Pi}^2[J', J] \begin{pmatrix} J' & 1 & J \\ 0 & 1 & -1 \end{pmatrix}^2 \begin{pmatrix} J' & 1 & J \\ -M' & q & M \end{pmatrix}^2, \tag{8.57}$$

where

$$\mu_{\Sigma\Pi} = \langle [^1\Sigma][v'] \mid d_{0-1}^{(01)} \mid [^1\Pi\ 1][v] \rangle. \tag{8.58}$$

Although the 3-j symbols in the expression (8.57) can be rapidly found from tables, it is useful to have explicit algebraic expressions for $\mathcal{S}_{ed}$. For the transition $J' \to J$, there are three cases to consider: $J' = J - 1$, $J' = J$, and $J' = J + 1$. In keeping with traditional nomenclature, they are labeled by P, Q, and R respectively. It is also convenient to take the average of the strengths for the two circular polarizations defined by $q = \pm 1$. This gives the strength for either of the two Cartesian components π_x or π_y, while the component for which $q = 0$ corresponds to π_z as it stands. The former are appropriate to linear polarizations perpendicular to the Zeeman field $\mathbf{H}$, the latter to a polarization parallel to $\mathbf{H}$. The labels $\parallel$ and $\perp$ are useful to make the distinction here. The various values of $\mathcal{S}_{ed}$ are calculated with the aid of the formulae of Appendix 1. The results are set out in Table 8.1, and the transitions to the lowest level of $^1\Pi$ are sketched in Fig. 8.1. If the various components of $^1\Sigma$ are populated with equal probability, and if the energies between the various components of $^1\Sigma$ and $^1\Pi$ are small compared to the transition quantum $\hbar\omega$, then Table 8.1 gives the relative in-

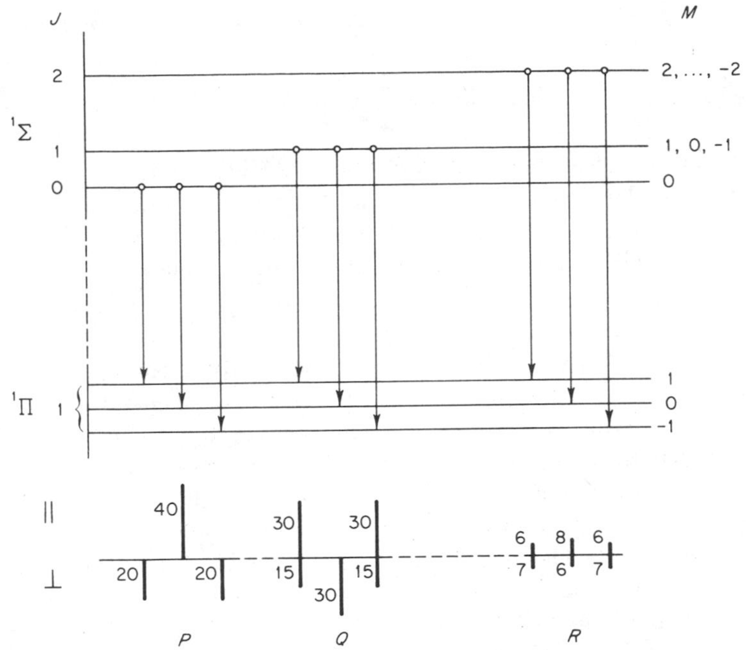

Fig. 8.1. Transitions and electric-dipole strengths for certain components of $^1\Sigma \rightarrow {}^1\Pi$ in the presence of a constant external magnetic field. This field splits the $J = 1$ level of $^1\Pi$ into its three components but leaves the levels of $^1\Sigma$ intact. The relative strengths of the transitions are given by the numbered lengths of the lines in the lower part of the diagram. The patterns for transitions to other J levels of $^1\Pi$ have been reproduced by Herzberg [127, p. 302] from Crawford's original work [126].

tensities for spontaneous transitions. If, in addition, the frequency distribution of an applied electromagnetic field varies slowly with ω, then the entries also give relative intensities for induced transitions.

The succession of transitions of the types P, Q or R constitutes the three *branches* of a *band*, whose component lines (in the absence of an external magnetic field) occur at the angular frequencies ω determined by

$$\hbar\omega = A + BJ(J + 1) - B'J'(J' + 1), \tag{8.59}$$

where B and B' are two rotational constants. Like A, these depend on the electronic and vibrational descriptions of the upper and lower levels. The substitutions $J' = J$ or $J' = J \pm 1$ give a linear dependence of $\hbar\omega$ on J when $B' = B$, but any deviation from this equality permits the quadratic term in J to predominate for sufficiently large J. This parabolic dependence is a highly characteristic feature of molecular spectra.

We turn now to the transition moment $\mu_{\Sigma\Pi}$. With C as the origin of coordinates, we have

$$d_{0-1}^{(01)} = -e \sum_t \sqrt{\tfrac{1}{2}}\,(\xi_t - i\eta_t).$$

We expect the electronic matrix element $\langle {}^1\Sigma \mid d_{0-1}^{(01)} \mid {}^1\Pi\ 1 \rangle$ to be rather weakly dependent on the internuclear separation r, since $d_{0-1}^{(01)}$ is independent of this quantity. Assuming, then, that the electronic orbitals vary only slightly with r, we arrive at the approximate equation

$$\mu_{\Sigma\Pi} = \langle v \mid v' \rangle \langle {}^1\Sigma \mid d_{0-1}^{(01)} \mid {}^1\Pi\ 1 \rangle.$$

Thus the transition strength depends on the square of the overlap between $\langle v \mid$ and $\mid v' \rangle$. To a first approximation, these are the eigenstates of a simple-harmonic oscillator—or, more precisely, two simple-harmonic oscillators characterized by two potential functions $W(r)$ (given in Eq. (6.37)). A qualitative understanding of the dependence of intensities on vibrational states may be readily gained if these functions are known. For large quantum numbers v, the classical situation is obtained in which the oscillator is most likely to be bound at an extreme position; and these limiting positions often determine the regions of maximum overlap and hence the most favored transitions. Criteria of this kind constitute the so-called Franck–Condon principle. Of course, if both the initial and final states of a transition correspond to the same electronic state, then the oscillator functions derive from the same potential $W(r)$. In this case the overlap vanishes unless $v = v'$; so the transitions between different vibrational states can only occur in virtue of the term in $\mathbf{d}$ that involves $\mathbf{r}_{AB}$. This point is considered in a little more detail in Problem 8.5.

8.7 LINE STRENGTHS

The transition ${}^1\Sigma \to {}^1\Pi$ corresponds to $S = 0$, and is thus exceptionally simple to analyze. But even in the presence of spin, the situation for Hund's case (a) is scarcely more complicated. The strength for the electric-dipole transitions between two states of the type (6.63) is given either by zero (if their parities are the same) or by

$$\mathbb{S}_{\text{ed}} = \mid \langle [\nu\Lambda S\Sigma][v][JM_J\ -\Omega] \mid d_q \mid [\nu'\Lambda' S\Sigma][v'][J'M_J'\ -\Omega'] \rangle \mid^2.$$

The dependence of this expression on M_J, q, and M_J' is irrelevant when there are no external fields, since the Stark or Zeeman patterns collapse into single lines. It is appropriate to sum over M_J, q, and M_J' with the aid of the formula

$$\sum_{abc} \begin{pmatrix} A & B & C \\ a & b & c \end{pmatrix}^2 = 1.$$

The relevant sums over $\mathcal{S}_{ed}$ define the *line strengths* $\mathcal{L}_{ed}$. For the example in hand, we soon find that

$$\mathcal{L}_{ed} = \mu^2[J', J] \begin{pmatrix} J & 1 & J' \\ \Omega & \Omega' - \Omega & -\Omega' \end{pmatrix}^2, \tag{8.60}$$

where

$$\mu = \langle [\nu\Lambda S][v] \mid d_{0,\Lambda-\Lambda'}{}^{(01)} \mid [\nu'\Lambda' S][v'] \rangle.$$

The line strengths specify the crucial dependence on J and J' of the transition probabilities.

For Hund's case (b), we start with eigenstates of the type given in Eq. (6.64). In this case,

$$\begin{aligned} \mathcal{L}_{ed} &= | ([\nu\Lambda][v][N - \Lambda]SJ \,|{:}\, (\mathbf{D}^{(11)}\mathbf{d}^{(01)})^{(10)} \,{:}|\, [\nu'\Lambda'][v'][N' - \Lambda']SJ') |^2 \\ &= \frac{1}{3}[J, J'] \begin{Bmatrix} N & 1 & N' \\ J' & S & J \end{Bmatrix}^2 \\ &\quad \times | ([\nu\Lambda][v][N - \Lambda] \,|{:}\, D_{\cdot\Lambda'-\Lambda}{}^{(11)} d_{0,\Lambda-\Lambda'}{}^{(01)} \,{:}|\, [\nu'\Lambda'][v'][N' - \Lambda']) |^2 \\ &= \mu^2[N, J, N', J'] \begin{Bmatrix} N & 1 & N' \\ J' & S & J \end{Bmatrix}^2 \begin{pmatrix} N & 1 & N' \\ \Lambda & \Lambda' - \Lambda & -\Lambda' \end{pmatrix}^2, \end{aligned} \tag{8.61}$$

and the dependence of $\mathcal{L}_{ed}$ on N, J, N', and J' is obtained. Strictly speaking, a factor of 2 should be included on the right-hand sides of Eqs. (8.60)–(8.61) when the two kets making up the basic states $| a\pm \rangle$ or $| b\pm \rangle$ (given in Eqs. (6.63)–(6.64)) coalesce; but this doubling in strength is exactly balanced by the fact that the lines whose strengths we have been calculating in this section occur in pairs. For between the two states $| a\pm \rangle$ and the two states $| a'\pm \rangle$, parity considerations allow two equal transitions and forbid two others.

Explicit expressions for the line strengths $\mathcal{L}_{ed}$ can be obtained by replacing the 3-j and 6-j symbols in Eqs. (8.60)–(8.61) by their detailed algebraic forms. An extensive tabulation has been made by Kovacs [90]. His expressions for the line strengths correspond to our $\mu^{-2}\mathcal{L}_{ed}$. Dimensionless quantities of this kind are sometimes called Hönl–London factors (see, for example, Tatum [128]). From Eq. (8.60), we find

$$\sum_J \mu^{-2}\mathcal{L}_{ed} = 2J' + 1;$$

the same result is obtained from Eq. (8.61) if the sum over J is extended to a double sum over J and N. These sum rules remain valid if primed quantities are replaced by unprimed quantities and vice versa.

8.8 MAGNETIC-DIPOLE AND QUADRUPOLE RADIATION

The methods that have been applied to electric-dipole radiation can be easily extended to cope with V_{md} and V_{eq} of Eqs. (8.52). The formal transference of the origin of the plane-wave expansion to C, the center of mass of the nuclear frame, produces additional terms in the interaction Hamiltonian. However, since both V_{md} and V_{eq} are quadratic functions of a particle's coordinates, the substitutions

$$\mathbf{r}_t = \mathbf{r}_{tC} + \mathbf{r}_C, \qquad \mathbf{r}_X = \mathbf{r}_{XC} + \mathbf{r}_C$$

add either a constant term (which possesses null off-diagonal matrix elements) or a term linear in the coordinates. This last merely leads to a small correction to the electric-dipole part of the expansion. Hence we can take C as the origin for the crucial tensors $\mathbf{L} + 2\mathbf{S}$ and $\mathbf{Q}^{(2)}$. The quadratic character of V_{md} and V_{eq} immediately leads to the selection rule that all transitions can only take place between states of identical parity. In those cases where the intrinsic parity labels g and u can also be used, we can deduce in the same way that the only allowed transitions are of the type $g \rightarrow g$ and $u \rightarrow u$.

Further selection rules are obtained from considerations of angular momentum. Let us take magnetic-dipole radiation first. Although V_{md} depends on spin, it does so through the total spin S itself. So we have

$$\Delta S = 0, \qquad \Delta M_S = 0, \pm 1, \qquad \Delta \Sigma = 0, \pm 1.$$

From the properties of $\mathbf{D}^{(11)}$, all the rules (8.54) for electric-dipole radiation are obtained again, as well as the elimination of the branches for which $\Delta J = 0$ or $\Delta N = 0$ when $\Omega = 0$ (for Hund's case (a)) or $\Lambda = 0$ (for Hund's case (b)). However, a new rule appears when we make the replacement

$$d_{00}{}^{(01)} \rightarrow L_{00}{}^{(01)} + S_{00}{}^{(01)}.$$

The latter, which is identical to $L_\zeta + 2S_\zeta$, possesses null matrix elements when set between a pair of *distinct* Σ states, as well as when set between *any* pair of $^1\Sigma$ states. In other words, the following transitions are *forbidden*:

$$[\nu\Sigma] \rightarrow [\nu'\Sigma] \qquad (\nu' \neq \nu)$$

$$^1\Sigma \rightarrow {}^1\Sigma \tag{8.62}$$

On the other hand, magnetic-dipole transitions within the levels of a state such as ${}^3\Sigma_g^-$ are permitted, and, indeed, have been observed in the microwave spectrum of the ground state of O_2, for example [129, 130].

The extensions to electric-quadrupole radiation are quite straightforward, the main difference being that $\mathbf{Q}^{(2)}$ is a tensor of rank 2 rather than a vector. Thus we have

$$\Delta\Xi = 0, \pm 1, \pm 2 \qquad (\Xi \equiv J, \Omega, M_J, N, \Lambda, M), \tag{8.63}$$

with the additional restriction that the triads $(J, 2, J')$ or $(N, 2, N')$ must satisfy the triangular condition. It is now the turn of the branches $\Delta J = \pm 1$ or $\Delta N = \pm 1$ to disappear when $\Omega = \Omega' = 0$ or $\Lambda = \Lambda' = 0$ respectively. Since $\mathbf{Q}^{(2)}$ is independent of spin,

$$\Delta S = \Delta M_S = \Delta\Sigma = 0.$$

The component $Q_{00}{}^{(02)}$ is invariant under the reflection operation R, from which we can deduce that electric-quadrupole radiation is forbidden for $\Sigma^+ \leftrightarrow \Sigma^-$. (The use of R yields no new selection rules for magnetic-dipole radiation.)

Turning now to the calculation of transition strengths, we at once see that the analysis for magnetic-dipole radiation is very similar to that for electric-dipole radiation. The relevant vector of the electromagnetic field is **H** rather than **E**, so the labels $\|$ and $\perp$ attached to the Zeeman patterns in Fig. 8.1 must be reinterpreted to describe how **H** stands with respect to the direction of the external constant magnetic field; but the relative transition strengths given in the figure remain valid. This is because both **d** and $\mathbf{L} + 2\mathbf{S}$ are vectors: a distinction between them appears only when the purely electronic matrix elements of $d_{00}{}^{(01)}$ and $L_\zeta + 2S_\zeta$ are compared. Thus Eqs. (8.60)–(8.61) for the line strengths $\mathfrak{L}_{\text{ed}}$ can be immediately taken over to give expressions for $\mathfrak{L}_{\text{md}}$. The only change we have to make is to replace $\mathbf{d}^{(01)}$ by $\mathbf{L}^{(01)} + 2\mathbf{S}^{(01)}$ in the definition of μ.

An an example of the calculation of line strengths for electric-quadrupole radiation, we take the case of ${}^1\Sigma v \rightarrow {}^1\Sigma v'$. Some extremely feeble transitions of this kind have been observed by Herzberg [131] within the ground electronic state ${}^1\Sigma_g^+$ of H_2. Applying the WE theorem to a component $Q_q{}^{(2)}$ of $\mathbf{Q}^{(2)}$, and summing over the magnetic quantum numbers, we get

$$\mathfrak{L}_{\text{eq}} = \frac{1}{6} \,|\, ([{}^1\Sigma v][N0] \,|\!:\, (\mathbf{D}^{(22)}\mathbf{Q}^{(02)})^{(20)} \,:\!|\, [{}^1\Sigma v'][N'0]) \,|^2$$

$$= \frac{1}{6}[N, N'] \begin{pmatrix} N & 2 & N' \\ 0 & 0 & 0 \end{pmatrix}^2 |\, \langle {}^1\Sigma v \,|\, Q_{00}{}^{(02)} \,|\, {}^1\Sigma v' \rangle \,|^2.$$

As expected, the conditions for the nonvanishing of the 3-j symbol lead to the selection rule $\Delta N = 0, \pm 2$. Thus three branches are anticipated. The fact that the transitions take place within a single electronic state permits an approximate selection rule on v to be established. Since $v \neq v'$, the first term on the right-hand side of Eq. (8.49) can only contribute to the matrix element of $Q_{00}{}^{(02)}$ if significant variations in the purely electronic quadrupole moment of the molecule occur in the range of r (the internuclear distance) where the overlap between $\langle v \mid$ and $\mid v' \rangle$ is appreciable. We would thus expect the second term of Eq. (8.49) to make the major contribution to the matrix element; and, if this is the case, we can write

$$Q_{00}{}^{(02)} = eZ_A r_{AC}{}^2 + eZ_B r_{BC}{}^2 = er^2(Z_A M_B{}^2 + Z_B M_A{}^2)/(M_A + M_B)^2.$$

Thus

$$\mathfrak{L}_{\text{eq}} = \frac{1}{6}[N, N'] \begin{pmatrix} N & 2 & N' \\ 0 & 0 & 0 \end{pmatrix}^2 \frac{e^2(Z_A M_B{}^2 + Z_B M_A{}^2)^2}{(M_A + M_B)^4} \mid \langle v \mid r^2 \mid v' \rangle \mid^2. \tag{8.64}$$

We now make the substitution

$$r^2 = (r - r_e)^2 + 2(r - r_e) + r_e{}^2, \tag{8.65}$$

thereby introducing the variable $r - r_e$ in terms of which the oscillator eigenfunctions (6.39) are defined. The elementary properties of these functions yield

$$\Delta v = 0, \pm 2 \qquad \text{for} \quad (r - r_e)^2,$$

$$\Delta v = \pm 1 \qquad \text{for} \quad r_e(r - r_e),$$

$$\Delta v = 0 \qquad \text{for } r_e{}^2.$$

So the selection rule on r^2 is

$$\Delta v = 0, \pm 1, \pm 2.$$

Some of the lines reported by Herzberg [131] correspond to $\Delta v = 3$, a result that indicates that the assumption of an invariant electronic quadrupole moment cannot be entirely justified.

8.9 CLASSICAL LIMITS

Our concern to this point has been primarily with low-lying or metastable molecular levels characterized by small angular momentum quantum numbers. However, the advent of the laser has made it possible to study in detail transitions to rotational levels for which $J > 100$. This makes it of considerable interest to analyze the passage from quantum mechanics to classical mechanics. The behavior of a 6-j symbol when some of its arguments become large has been described by Ponzano and Regge [132]. Relevant though this work is, considerable insight into the approach to the classical limit can be obtained without going through a detailed analysis. As an example, we take the hyperfine structure of a line in the spectrum of $^{127}I_2$ corresponding to the transition

$$[{}^1\Sigma_{0g}{}^+, v = 21, J = 117] \rightarrow [{}^3\Pi_{0u}, v = 1, J = 116].$$

This has been observed by Hänsch *et al.* [133]. Spin–orbit interaction mixes states of different S and is responsible for the apparent breakdown in the selection rule $\Delta S = 0$. There is added interest for us in the hyperfine structure because each nucleus ^{127}I (for which $I = 5/2$) possesses an electric quadrupole moment, and it is the interaction of these moments with the electronic motion, rather than the magnetic-dipolar interaction of Section 7.8, which determines the general features of the hyperfine structure.

To allow for the existence of nuclear quadrupole moments, we relax the condition of point nuclei, and we interpret $\mathbf{r}_X$ as defining the position of the center of gravity of nucleus X. Each proton p of nucleus X interacts with an electron t through a term of the type $-e^2/r_{tp}$, which we expand under the assumption $r_{pX} \ll r_{tX}$:

$$\frac{e^2}{r_{tp}} = -\frac{e^2}{|\mathbf{r}_{tX} - \mathbf{r}_{pX}|} = -e^2 \sum_k \frac{r_{pX}{}^k}{r_{tX}{}^{k+1}} \mathbf{C}^{(k)}(\theta_{pX}\phi_{pX}) \cdot \mathbf{C}^{(k)}(\theta_{tX}\phi_{tX}). \qquad (8.66)$$

The leading term $-e^2/r_{tX}$ gives the result corresponding to a point nucleus; the term for which $k = 1$ has null matrix elements provided the nucleus exists in a state of well-defined parity with respect to inversion in its center of mass; and the next term (the last we need consider) represents the contribution to the quadrupole hyperfine interaction H_{qh} corresponding to one particular triad (ptX).

For present purposes we need not study in detail the development of the $k = 2$ term in Eq. (8.66). A double application of the WE theorem in the nuclear and molecular spaces permits us to write, for two identical nuclei,

$$H_{\mathrm{qh}} = B \sum_X (\mathbf{I}_X\mathbf{I}_X)^{(2)} \cdot (\mathbf{JJ})^{(2)}, \qquad (8.67)$$

provided our interest lies in matrix elements diagonal with respect to I and J. Application of Eqs. (1.65)–(1.67) gives us

$$\langle J(II)TF \mid H_{\mathrm{qh}} \mid J(II)T'F\rangle$$
$$= B[(-1)^{J+F+2I} + (-1)^{J+F+2I+T+T'}][T, T']^{1/2}$$
$$\times \begin{Bmatrix} T & 2 & T' \\ J & F & J \end{Bmatrix} \begin{Bmatrix} T & 2 & T' \\ I & I & I \end{Bmatrix} (J \,\|\, (\mathbf{JJ})^{(2)} \,\|\, J)(I \,\|\, (\mathbf{II})^{(2)} \,\|\, I). \qquad (8.68)$$

The two-part phase factor arises from the two terms in the sum over X. Although it indicates that T and T' must both be odd or both be even for the matrix element of H_{qh} not to vanish, no other case arises in practice. To see why this should be so, we note that Eq. (6.67) indicates that the eigenvalue of the nuclear interchange operator O is $(-1)^J$ for ${}^1\Sigma_g{}^+$; and this must be combined with $(-1)^{2I-T}$ (from Eq. 6.69) to give -1, since, for ^{127}I, the nuclear spin I is half-integral. Thus, if J is even, the acceptable values of T are 0, 2, 4; and if J is odd, they are 1, 3, 5. This result applies to ${}^1\Sigma_g{}^+$. In the case of a J' level of ${}^3\Pi_{0u}$, we first note that the parity, from Eq. (6.63), is $(-1)^{J'+1\pm1}$ for $|\,a\pm\rangle$; and, for an electric-dipole transition to take place, this must be opposite to the parity of the J level of ${}^1\Sigma_{0g}{}^+$, namely $(-1)^J$. Since $\Delta J = 1$ for the transition under investigation, the state $|\,a-\rangle$ must be selected. Equation (6.65) now gives $(-1)^{J'+1}$ as the eigenvalue of O; and we conclude that the acceptable values of T are 1, 3, 5 when J' is even and 0, 2, 4 when it is odd. For $J = 117$ and $J' = 116$, the values of T we need consider (for both the initial and the final levels) are therefore 1, 3, 5.

The reduced matrix elements in Eq. (8.68) can be readily evaluated. Thus we find

$$(J\,||\,(\mathbf{JJ})^{(2)}\,||\,J) = \{J(J+1)(2J-1)(2J+1)(2J+3)/6\}^{1/2},$$

and a similar expression for $(I\,||\,(\mathbf{II})^{(2)}\,||\,I)$. The matrices for H_{qh} can now be set up (for a given J and I) in terms of a single parameter B. The largest matrix possesses three rows and three columns. Diagonalization is a straightforward affair, and 21 roots occur. However, in the limit $J\to\infty$, these 21 roots collapse to just 6 distinct values, and this indicates that a more suitable set of basis states must exist in the classical limit. Some inkling of what this might be can be obtained by picking a component of an F level for which $M_F = F$. An expansion in uncoupled states runs

$$|\,(JT)FF\rangle = a\,|\,JJ,\,TM_T\rangle + b\,|\,J\,J-1,\,T\,M_T+1\rangle + \cdots, \qquad (8.69)$$

where $M_T = F - J$ and the coefficients $a, b, \ldots$ are CG coefficients. Their values may be obtained by operating on the right-hand side of Eq. (8.69) with F_+ $(= J_+ + T_+)$ and requiring that the result be zero (since M_F has its maximum value). From Eqs. (1.3) we get a succession of equations of the type

$$a\{T(T+1) - M_T(M_T+1)\}^{1/2} + b\{J(J+1) - (J-1)J\}^{1/2} = 0.$$

which indicate that

$$b \sim -aJ^{-1/2}, \qquad c \sim -bJ^{-1/2}, \ldots .$$

So, in the limit of large J,

$$|(JT)FF\rangle = |JJ, TM_T\rangle \qquad (M_T = F - J).$$

We can uncouple the nuclear part and take the states

$$|JJ, IM_{IA}, IM_{IB}\rangle \tag{8.70}$$

as a basis. These are diagonal with respect to H_{qh}; in fact

$$\begin{aligned}\langle JJ, IM_{IA}, IM_{IB} | H_{\mathrm{qh}} | JJ, IM_{IA}, IM_{IB}\rangle &= B\langle JJ | (\mathbf{JJ})_0^{(2)} | JJ\rangle[\langle IM_{IA} | (\mathbf{II})_0^{(2)} | IM_{IA}\rangle \\ &\qquad + \langle IM_{IB} | (\mathbf{II})_0^{(2)} | IM_{IB}\rangle] \\ &= \tfrac{1}{6}BJ(2J-1)[\{3M_{IA}^2 - I(I+1)\} + \{3M_{IB}^2 - I(I+1)\}].\end{aligned} \tag{8.71}$$

As M_{IA} and M_{IB} run in integral steps within the limits $\pm 5/2$, the expression (8.71) assumes just six distinct values. The magnitude of F corresponding to a given M_{IA} and M_{IB} is determined by

$$F = J + M_T = J + M_{IA} + M_{IB}.$$

In the classical limit, the axes of **F** and **J** virtually coincide, and the energy spectrum corresponds to two independent quadrupoles in an axial field.

The basis (8.70) is very suitable for calculating the effect of the dipolar part H_{dh} of the hyperfine interaction. We have

$$H_{\mathrm{dh}} = H_{\mathrm{no}} + H_{\mathrm{ns}} + H_{\mathrm{nsc}},$$

where the terms on the right are defined in Section 7.8. For matrix elements diagonal with respect to J and I, it is convenient to use the equivalence

$$H_{\mathrm{dh}} \equiv A \sum_X \mathbf{I}_X \cdot \mathbf{J}. \tag{8.72}$$

We now have

$$\langle JJ, IM_{IA}, IM_{IB} | H_{\mathrm{dh}} | JJ, IM_{IA}, IM_{IB}\rangle = AJ(M_{IA} + M_{IB}) = AJM_T.$$

The resulting splittings of the six degenerate quadrupole levels are sketched in Fig. 8.2 for $A \ll B$.

The hyperfine splittings of the $J = 117$ level of $^1\Sigma_{0g}{}^+$ and the $J = 116$ level of $^3\Pi_{0u}$ are similar to the energy-level scheme shown on the left in Fig. 8.2, though the A and B values are expected to be different for the two levels. To determine the hyperfine structure of the line connecting them, we need to know the selection rules for transitions between states of the type (8.70). Since electric-dipole radiation does not involve the internal

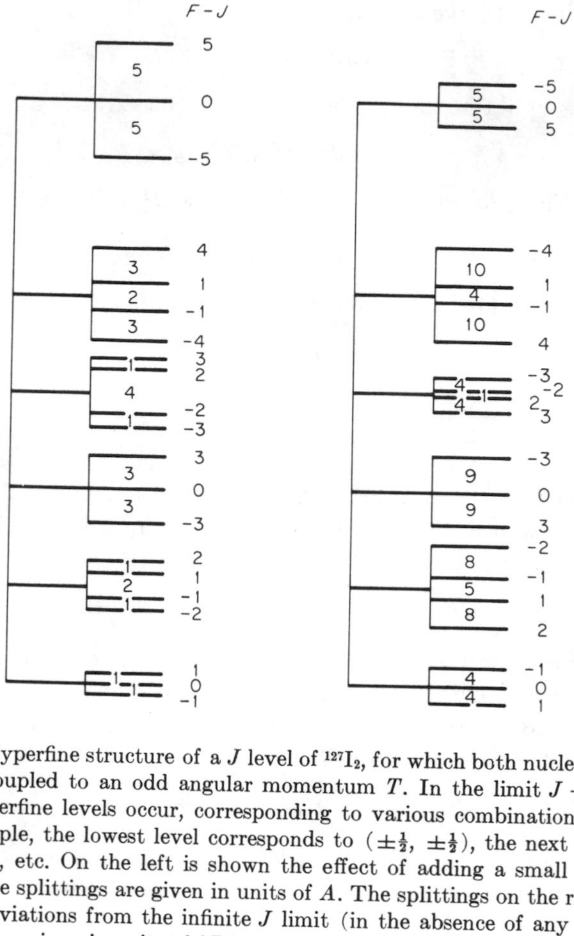

Fig. 8.2. Hyperfine structure of a J level of $^{127}I_2$, for which both nuclei possess spins of $\frac{5}{2}$ and are coupled to an odd angular momentum T. In the limit $J \to \infty$, just six degenerate hyperfine levels occur, corresponding to various combinations of M_{IA} and M_{IB}. For example, the lowest level corresponds to $(\pm\frac{1}{2}, \pm\frac{1}{2})$, the next to $(\pm\frac{1}{2}, \pm\frac{3}{2})$ and $(\pm\frac{3}{2}, \pm\frac{1}{2})$, etc. On the left is shown the effect of adding a small dipolar interaction $A\mathbf{I}\cdot\mathbf{J}$; the splittings are given in units of A. The splittings on the right represent the effect of deviations from the infinite J limit (in the absence of any dipolar interactions); they are given in units of $\frac{1}{3}B(2J - 1)$.

coordinates of the nuclei, we have at once

$$\Delta M_{IA} = \Delta M_{IB} = 0.$$

Thus transitions can only take place between corresponding hyperfine components, and so the hyperfine structure of the line resembles the hyperfine splitting of a level. The absence of cross connections means, however, that the observed structure of the line can only yield the differences between the hyperfine splittings rather than their actual values.

It is of some interest to examine small departures from the classical limit. These are sketched on the right of Fig. 8.2 in the absence of dipolar interactions. If small, the latter can be included simply by adding in their contributions.

PROBLEMS

8.1 A heteronuclear diatomic molecule, for which both nuclei possess zero spin, is subjected to a small constant electric field E. A pair of rotational levels characterized, for Hund's case (b), by identical quantum numbers $[\nu\ \Lambda\ vNSJ]$ but opposite parity are separated by an energy 2δ when $E = 0$. Prove that the energies W of their components are given by

$$W = \pm(\delta^2 + \epsilon^2)^{1/2},$$

where ϵ is related to the electric dipole moment μ_e of the molecule in the state $[\nu\Lambda v]$ by the equation

$$\epsilon = E\mu_e M_J \Lambda \frac{N(N+1) + J(J+1) - S(S+1)}{2J(J+1)N(N+1)},$$

provided that the energies between levels with different J are large compared to δ.

8.2 Show that the partial Zeeman Hamiltonian H_Z', defined in Eq. (8.22), possesses diagonal matrix elements with respect to the Hund's case (a) states

$$|\,[\nu\Lambda S\Sigma][v][JM_J\ -\Omega]\rangle$$

given by

$$\tfrac{1}{2}\beta\alpha^2 a_0 H[\Omega M_J/J(J+1)]\langle[\nu\Lambda S\Sigma][v]\,|\,U\,|\,[\nu\Lambda S\Sigma][v]\rangle,$$

where the operator U is defined in terms of the eigenvalue σ of the projection s_ζ of $\mathbf{s}$ by

$$U = \sum_{tX} Z_X r_{tX}^{-3}\sigma_t(\xi_t^2 + \eta_t^2),$$

where $\xi_t^2 + \eta_t^2$ is the square of the distance of electron t from the internuclear axis.

8.3 The orbit–orbit interaction H_{oo} between nucleus X and electron t of a diatomic molecule can be written as [113]

$$H_{oo} = 2\beta\beta_N(Z_X/M_X)M_P[r_{Xt}^{-1}\mathbf{p}_X\cdot\mathbf{p}_t + r_{Xt}^{-3}\mathbf{r}_{Xt}\cdot(\mathbf{r}_{Xt}\cdot\mathbf{p}_X)\mathbf{p}_t].$$

where $\hbar\mathbf{p}_i$ is the linear momentum of particle i. Show that, when the molecule is subjected to a constant magnetic field $\mathbf{H}$, the orbit–orbit interaction gives rise to a contribution to the Zeeman Hamiltonian of

$$\beta_N \alpha^2 \mathbf{H} \cdot (\mathbf{Y} + \mathbf{Z}),$$

where α is the fine-structure constant and $\mathbf{Y}$ is given by

$$\mathbf{Y} = \tfrac{1}{2} a_0 M_P \sum_{tX} (Z_X / M_X r_{Xt}) [(\mathbf{r}_t \times \mathbf{p}_X) + r_{Xt}^{-2} (\mathbf{r}_t \times \mathbf{r}_{Xt}) (\mathbf{r}_{Xt} \cdot \mathbf{p}_X)],$$

with $\mathbf{Z}$ defined analogously for those terms involving the electrons' momenta. Prove that, for matrix elements diagonal with respect to the quantum numbers $[\nu \Lambda S M_S, v, N]$, the operator equivalence

$$\mathbf{H} \cdot \mathbf{Y} \equiv \gamma \mathbf{H} \cdot \mathbf{R}$$

is valid, where $\mathbf{R}$ is the nuclear angular momentum operator $\boldsymbol{l}_{BA}$. Express γ as

$$\gamma = \langle [\nu\Lambda][v] \mid T_{00}^{(01)} \mid [\nu\Lambda][v] \rangle,$$

where

$$\mathbf{T}^{(01)} = \tfrac{1}{6} a_0 M_P r_{AB}^{-1} \sum_t [(Z_B / M_B r_{Bt}) \{ (5/2)^{1/2} (\mathbf{r}_{tC}^{(01)} \mathbf{C}_{Bt}^{(02)})^{(01)} + 4 \mathbf{r}_{tC}^{(01)} \}$$
$$- (Z_A / M_A r_{At}) \{ (5/2)^{1/2} (\mathbf{r}_{tC}^{(01)} \mathbf{C}_{At}^{(02)})^{(01)}$$
$$+ 4 \mathbf{r}_{tC}^{(01)} \}].$$

Check the result by taking the special case when an electron happens to lie on the internuclear axis.

8.4 A tensor $\mathbf{T}^{(l)}$ is a function of $\mathbf{r}$. Two other tensors $\mathbf{U}^{(\kappa)}$ and $\mathbf{V}^{(l)}$ are constructed from the derivatives of $\mathbf{T}^{(l)}$ by writing

$$\mathbf{U}^{(\kappa)} = (\boldsymbol{\nabla}^{(1)} \mathbf{T}^{(l)})^{(\kappa)}, \qquad \mathbf{V}^{(l)} = \boldsymbol{\nabla}^2 \mathbf{T}^{(l)}.$$

Prove

$$[\boldsymbol{\nabla}^2, \mathbf{T}^{(l)}] = \mathbf{V}^{(l)} - 2 \sum_{\kappa} \{ (2\kappa + 1)/(2l + 1) \}^{1/2} (\mathbf{U}^{(\kappa)} \boldsymbol{\nabla}^{(1)})^{(l)}.$$

Take $\mathbf{T}^{(l)} = r^l \mathbf{C}^{(l)}$ and show that

$$\mathbf{V}^{(l)} = \mathbf{U}^{(l)} = \mathbf{U}^{(l+1)} = 0,$$
$$\mathbf{U}^{(l-1)} = - \{ l(2l + 1) \}^{1/2} \mathbf{T}^{(l-1)}.$$

Prove that, when set between an eigenbra and eigenket of the total nonrelativistic Hamiltonian of a diatomic molecule (corresponding to eigenvalues E_a and E_b for which $E_a - E_b = \hbar kc$), the equivalence

$$\hbar (\mathbf{T}^{(l-1)} \boldsymbol{\nabla}^{(1)})^{(l)} \equiv -kMc \{ l(2l - 1) \}^{-1/2} \mathbf{T}^{(l)}$$

is valid, where M is the mass of that particle for which $-i\boldsymbol{\nabla}$ is the momentum.

8.5 Two vibrational states $|v\rangle$ and $|v'\rangle$ belong to the same electronic state of a diatomic molecule. Use Eq. (8.8) for the dipole-moment operator $\mathbf{d}$ to show that electric-dipole transitions can only occur for $v - v' = \pm 1$, provided it is a good approximation to assume that the integral over the electrons' coordinates is independent of the internuclear separation r. With the additional assumption that the rotational structure can be neglected, show that the spectrum consists of a single line provided the vibrational motion is simple-harmonic. Prove that this line does not occur for homonuclear diatomics.

8.6 The Raman effect is a second order radiative process in which an incident wave (characterized by a polarization vector $\boldsymbol{\pi}$), impinging on a diatomic molecule, produces a scattered wave (characterized by $\boldsymbol{\pi}'$) and excites (or de-excites) the molecule from $|a\rangle$ to $|b\rangle$. An excited state $|c\rangle$ of the molecule contributes to the strength of the scattered light through various combinations of products of the type

$$\langle a \mid \boldsymbol{\pi}\cdot\mathbf{d} \mid c\rangle\langle c \mid \boldsymbol{\pi}'\cdot\mathbf{d} \mid b\rangle.$$

Express $\mathbf{d}^{(10)}$ in terms of $\mathbf{D}^{(11)}$ and $\mathbf{d}^{(01)}$, thereby obtaining the selection rules

$$\Delta J = 0, \pm 1, \pm 2 \qquad \text{(Hund's case (a))}$$

and

$$\Delta N = 0, \pm 1, \pm 2 \qquad \text{(Hund's case (b))}.$$

P_m is defined by writing

$$P_m = \begin{pmatrix} J & 1 & J'' \\ 0 & -m & m \end{pmatrix}\begin{pmatrix} J'' & 1 & J' \\ -m & m & 0 \end{pmatrix}.$$

Prove that, for integers J and J' differing by 1, the equations

$$P_0 = 0, \qquad P_m + P_{-m} = 0$$

are valid. Deduce that, when the initial and final states $|a\rangle$ and $|b\rangle$ are both characterized (for Hund's case (a)) by $\Omega = 0$, the Raman bands corresponding to $\Delta J = \pm 1$ disappear. Show that the analogous extinction of the $\Delta N = \pm 1$ bands takes place (for Hund's case (b)) for states for which $\Lambda = 0$.

8.7 Construct the diagram corresponding to Fig. 8.2 when the total nuclear spin T is even rather than odd. Show that the new energy-level scheme corresponding to deviations from the classical limit could be derived from the old one (apart from F assignments) simply by erasing 10 hyperfine levels.

8.8 From Eq. (8.61), the selection rules for electric-dipole radiation for Hund's case (b) are seen to be

$$\Delta J = 0, \pm 1; \qquad \Delta N = 0, \pm 1.$$

Show that, in the classical limit when J and N become very large, the additional selection rule

$$\Delta J = \Delta N$$

applies.

8.9 A particle of charge q, mass M, and magnetic moment $\boldsymbol{\mu}$ interacts with an electromagnetic wave characterized by $\mathbf{A}_0$ and $\mathbf{k}$, in the notation of Section 8.5. Show that, in the long-wavelength limit, the interaction Hamiltonian of Eq. (8.35) can be written as

$$h_{\text{int}} = h_1^{\text{e}} + h_1^{\text{m}} + h_2^{\text{e}} + h_2^{\text{m}},$$

where

$$h_1^{\text{e}} = -q \sum_t i^t\{2^t t!/(2t)!\}\{(2t-1)/t\}^{1/2} k^t r^t (\mathbf{C}_k^{(t-1)}\mathbf{A}_0)^{(t)} \cdot \mathbf{C}^{(t)},$$

$$h_1^{\text{m}} = (q\hbar/Mc) \sum_t i^{t+1}\{2^t t!/(2t)!\}\{(2t-1)/(t+1)\}^{1/2}$$
$$\times k^t r^{t-1} (\mathbf{C}_k^{(t)}\mathbf{A}_0)^{(t)} \cdot (\mathbf{C}^{(t-1)}\boldsymbol{l})^{(t)},$$

$$h_2^{\text{e}} = -\sum_t i^t\{2^t(t-1)!/(2t)!\}\{t(2t+3)\}^{1/2} k^{t+1} r^t (\mathbf{C}_k^{(t+1)}\mathbf{A}_0)^{(t)} \cdot (\boldsymbol{\mu}\mathbf{C}^{(t)})^{(t)},$$

$$h_2^{\text{m}} = \sum_t i^{t+1}\{2^t t!/(2t)!\}\{(t+1)(2t-1)\}^{1/2} k^t r^{t-1} (\mathbf{C}_k^{(t)}\mathbf{A}_0)^{(t)} \cdot (\boldsymbol{\mu}\mathbf{C}^{(t-1)})^{(t)}.$$

In these expressions, the tensors $\mathbf{C}_k^{(x)}$ (where $x = t-1$, t, or $t+1$) are functions of the polar angles θ_k and ϕ_k that define the direction of $\mathbf{k}$. (Alternative expressions are given by Brink and Satchler [3] in their Eqs. (6.11) and (6.12).)

Prove that the terms for which $t = 1$ and $t = 2$ in the expression for h_1^{e} reduce to H_{ed} and H_{eq} (defined in Eqs. (8.41) and (8.48)). Show too that the leading terms in h_1^{m} and h_2^{m} combine to yield H_{md} (defined in Eq. (8.45)).

9

Perturbations

9.1 INTERMEDIATE COUPLING

All the energy-level calculations of the previous chapters have been limited to diagonal matrix elements. This supposes that diatomic molecules can be adequately described by basis states of the type $| a\pm \rangle$ (for Hund's case (a)) or $| b\pm \rangle$ (for Hund's case (b)). As might be anticipated, this is often not so. Many physical situations arise in which it is not possible to neglect off-diagonal matrix elements. The great variety of terms in the molecular Hamiltonian offers an embarrassingly rich field for study. Fortunately, the general approach usually amounts to a straightforward application of perturbation theory, and the nature of the calculations can be reasonably well conveyed by a few examples. No attempt is made here to give a comprehensive survey of second order effects.

A natural starting point is a discussion of the transition from Hund's case (a) to case (b). In question here is the extent to which $\mathbf{S}$ couples to the orbital motion of the electrons, whose classification $[\nu \mid \Lambda \mid]$ is the same in both limits. The case of $S = \frac{1}{2}$ was studied by Hill and Van Vleck in a classic article [134]. This specialization is a convenient one for us to follow, since the small S value eliminates all contributions from the spin–spin interaction H_{ss}. If, further, we suppose that the spin–rotation interaction H_{sr} is sufficiently small to be neglected (which is usually the case when $\Lambda \neq 0$), the problem reduces to a study of the competing effects of the spin-orbit interaction H_{so} (given in Eq. (7.1)) and the rotational Hamiltonian H_r (given in Eqs. (6.31)). It is immaterial whether the basis for Hund's case (a) or Hund's case (b) is used to set up the matrix for $H_r + H_{so}$. Since H_{so} has already been studied for case (b) in Section 7.3, it may be more useful to look at the situation for case (a).

The scalar products that make up H_{so} are of the type $\mathbf{u}_{vt} \cdot \mathbf{s}_t$. For case (a), it is convenient to expand these products in terms of their components in the molecular frame F'. Only $u_{00}{}^{(01)}$ can connect states characterized by identical $|\Lambda|$ values. Since $u_{00}{}^{(01)}$ changes sign under the reflection operation R (of Section 6.9), we have

$$\langle \nu\Lambda \mid (u_{00}{}^{(01)})_{vt} \mid \nu\Lambda \rangle = -\langle \nu -\Lambda \mid (u_{00}{}^{(01)})_{vt} \mid \nu -\Lambda \rangle.$$

For a given $|\Lambda|$, the matrix elements of $(u_{00}{}^{(01)})_{vt}$ are thus proportional to Λ. Each $u_{00}{}^{(01)}$ is associated with a component $(s_{00}{}^{(01)})_t$, whose eigenvalues, for a given S, are proportional to Σ. We can conclude that the matrix elements of H_{so} in Hund's case (a) are diagonal with numerical values $A\Lambda\Sigma$, where A is independent of S and the sign of Λ.

The rotational term can be written (see Section 6.8) as

$$H_r' = B(\mathbf{J} - \mathbf{P})^2 = B(\mathbf{J} - \mathbf{S} - \mathbf{L})^2,$$

where $B = \hbar^2/2\mu r$. For states diagonal with respect to $|\Lambda|$, it is equivalent to

$$B(\mathbf{J}^2 + \mathbf{S}^2 + L_\xi^2 + L_\eta^2 + L_\zeta^2 - 2J_\zeta L_\zeta + 2S_\zeta L_\zeta - 2\mathbf{S}\cdot\mathbf{J}). \quad (9.1)$$

Since $J_\zeta - S_\zeta - L_\zeta$ has a null eigenvalue for the states $|a\pm\rangle$, we can subtract $(J_\zeta - S_\zeta - L_\zeta)^2$ from the expression (9.1), thereby getting

$$B(\mathbf{J}^2 - J_\zeta^2 + \mathbf{S}^2 - S_\zeta^2 + L_\xi^2 + L_\eta^2) - 2B(S_\xi J_\xi + S_\eta J_\eta).$$

The first term possesses the diagonal matrix elements

$$B[J(J+1) - \Omega^2 + S(S+1) - \Sigma^2 + \langle L_\xi^2 + L_\eta^2 \rangle].$$

The second term contributes only to those off-diagonal matrix elements for which $\Delta\Sigma = -\Delta\Omega = \pm 1$. For example,

$$-2B\langle [\nu\Lambda S\Sigma][v][JM_J\,-\Omega] \mid (S_\xi J_\xi + S_\eta J_\eta) \mid [\nu\Lambda S\,\Sigma - 1][v][JM_J\,-\Omega + 1] \rangle$$

$$= B\{S(S+1) - \Sigma(\Sigma - 1)\}^{1/2}\{J(J+1) - \Omega(\Omega - 1)\}^{1/2}.$$

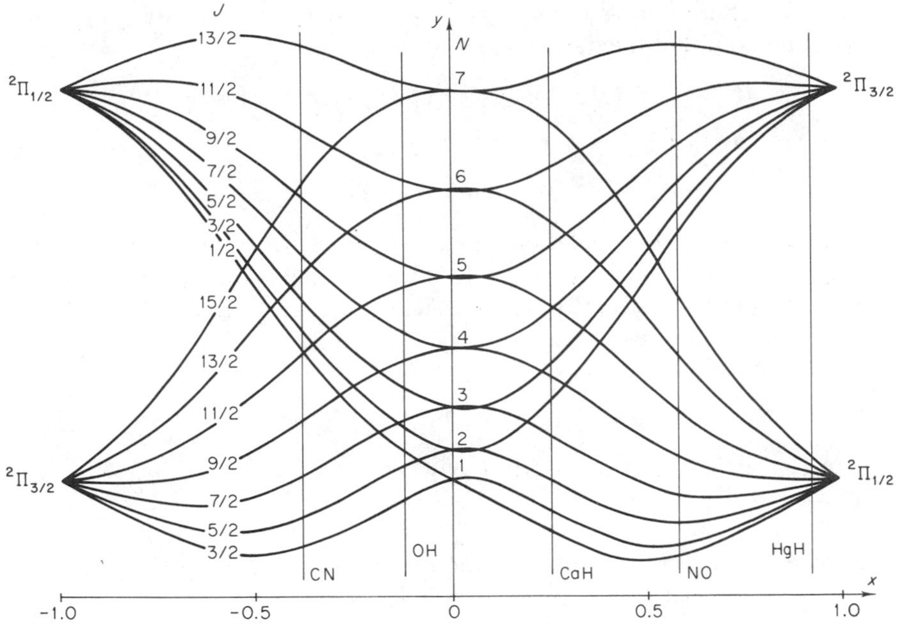

Fig. 9.1. Relative energies of $^2\Pi$ states in intermediate coupling between Hund's cases (a) and (b). The former corresponds either to the extreme right edge or the extreme left edge of the figure (depending on the sign of the spin–orbit coupling), the latter to the central axis labeled y. Only those rotational states corresponding to $1 \le N \le 7$ are shown. Ratios between energy separations are preserved in the figure, not absolute energies. Five experimentally observed $^2\Pi$ systems are specified by the intersections of the curves with the labeled straight lines. The appropriate values of x have been calculated either from the reported [127, p. 233; 135; 136] values of Y (which are converted to χ and fed directly into Eq. (9.3)), or else by referring immediately to the experimental data [137, 138].

The sign reversal here can be seen by noting that $-\mathbf{J}^{(01)}$ rather than $\mathbf{J}^{(01)}$ is an angular momentum vector, and $-\Omega$ in the bra is the eigenvalue of $-J_{00}{}^{(01)}$. (See also Problem 3.2.)

The matrix of $H_{so} + H_r$ can now be compiled. For $S = \frac{1}{2}$, the determinant of the matrix factorizes into a one-by-one determinant (for $J = |\Lambda| - \frac{1}{2}$) and a series of two-by-twos. Dropping the common diagonal contribution $\langle L_\xi{}^2 + L_\eta{}^2 \rangle$, we find that the determinant for arbitrary J ($> |\Lambda|$) runs

$$\begin{vmatrix} -\frac{1}{2}A\Lambda + B[(J+\frac{1}{2})^2 - \Lambda(\Lambda-1)] & B\{(J+\frac{1}{2})^2 - \Lambda^2\}^{1/2} \\ B\{(J+\frac{1}{2})^2 - \Lambda^2\}^{1/2} & \frac{1}{2}A\Lambda + B[(J+\frac{1}{2})^2 - \Lambda(\Lambda+1)] \end{vmatrix},$$

where the rows (and columns) are characterized by $\Omega = \Lambda - \frac{1}{2}$ and

$\Omega = \Lambda + \frac{1}{2}$. Subtracting the energy eigenvalue E from the diagonal elements and equating the determinant to zero, we get

$$E/B = (J + \tfrac{1}{2})^2 - \Lambda^2 \pm \{\tfrac{1}{4}Y(Y - 4)\Lambda^2 + (J + \tfrac{1}{2})^2\}^{1/2},$$

where $Y = A/B$. For $J = |\Lambda| - \frac{1}{2}$, we find

$$E/B = |\Lambda| (1 - \tfrac{1}{2}Y).$$

The properties of these solutions for E are most easily appreciated from a diagram. We specialize to $|\Lambda| = 1$, corresponding to $^2\Pi$ states. It is convenient to limit attention to the rotational levels for which $N < 8$. The energy-level structures in the limits of Hund's cases (a) and (b) can be matched by defining

$$y = [(E/B) - 28]/(1 + \chi^2)^{1/2}, \tag{9.2}$$

where $\chi = Y/54$. Plots of y against x, where

$$x = \chi/(1 + |\chi|), \tag{9.3}$$

are made in Fig. 9.1. The ratios of the relative energies are preserved, and a direct comparison with experiment is possible.

Extensions to other multiplicities are straightforward, though the ranks of the determinants increase as $2S + 1$. Kovacs [90] lists many of the matrix elements explicitly. The eigenfunctions obtained by diagonalizing the matrices are available for calculating smaller effects, such as those produced by H_{sr}. Of course, if $\Lambda = 0$, the product $A\Lambda\Sigma$ vanishes, and the effect of H_{sr} becomes unobscured. Unlike H_{so}, the spin–rotation interaction H_{sr} possesses the property that the splittings it produces in the rotational levels near the limit of Hund's case (b) increase with N, as has been mentioned in Section 7.6. Splittings of this kind in Σ states are sometimes referred to as ρ-doubling.

9.2 Λ-DOUBLING

In Sections 8.2 and 8.3, reference is made to the fact that the degenerate pairs of kets $|a\pm\rangle$ or $|b\pm\rangle$ of opposite parity often show small splittings. When $\Lambda \neq 0$, this effect is called Λ-doubling. To illustrate how it comes about, we confine ourselves to systems corresponding to Hund's case (a). The principal source of the doubling is the term

$$-\hbar^2(\mathbf{P}\cdot\mathbf{J})/\mu r^2, \tag{9.4}$$

which occurs in H_r' of Eq. (6.51). Only the part $P_\zeta J_\zeta$ of this Coriolis term plays a role in determining the zeroth-order states, and so the difference, which we define as

$$H_\Lambda = -\hbar^2(P_\xi J_\xi + P_\eta J_\eta)/\mu r^2, \tag{9.5}$$

must be treated as a perturbation. Since P_ξ and P_η can only connect states differing by at most one unit in Λ, we see that

$$\langle a\pm \mid H_\Lambda \mid a\pm \rangle = 0,$$

where any combination of signs can be taken. Thus there are no first order splittings.

To illustrate second-order effects, we take the $^3\Pi$ multiplet. Suppose the levels based on $^3\Pi_1$ are considered first. Since $^3\Pi_1$ corresponds to $\Omega = \pm 1$, the states of $^3\Pi$ can only be assigned $\Sigma = 0$ and $\Lambda = \pm 1$. As the perturbation H_Λ depends on spin through **S** itself, the selection rules for H_Λ are

$$\Delta S = 0, \qquad \Delta J = 0,$$

$$(\Delta\Sigma, \Delta\Lambda, \Delta\Omega) = (0, \pm 1, \mp 1) \quad \text{or} \quad (\pm 1, 0, \mp 1).$$

Thus the only states that can be directly coupled to $^3\Pi_1$ are those labeled $^3\Delta_2$, $^3\Pi_0$, $^3\Pi_2$, and $^3\Sigma_0$. The first three of these states (i.e., those for which $\Lambda \neq 0$) comprise doublets whose two components possess opposite parity. Any one such doublet can be represented by the pair of bras $\langle a'\pm \mid$, say. The interaction via H_Λ of this pair with the kets $\mid a\pm \rangle$ of $^3\Pi_1$ requires the evaluation of the four matrix elements

$$\langle a'\pm \mid H_\Lambda \mid a\pm \rangle. \tag{9.6}$$

Although it might be expected that the expansions (6.63) of $\langle a'\pm \mid$ and $\mid a\pm \rangle$ into their two components would convert (9.6) into the sum of four matrix elements, two of these four correspond to $\mid \Delta\Lambda \mid > 1$ and hence are necessarily zero. The remaining pair must be of equal magnitude in order to ensure that (9.6) vanishes when a bra and a ket of opposite parity are chosen. This implies that the two nonvanishing examples of (9.6) are of equal magnitude. We deduce that states of the type $^3\Delta_2$, $^3\Pi_0$, and $^3\Pi_2$ displace both components $\mid a\pm \rangle$ of $^3\Pi_1$ by an equal amount.

States assigned the labels $^3\Sigma_0$ do not occur as doublets of opposite parity, since the degeneracy corresponding to the interchange $(\Lambda\Sigma) \leftrightarrow (-\Lambda\ -\Sigma)$ is not present. The interaction of $^3\Pi_1$ with $^3\Sigma_0$ thus provides a source for Λ-doubling. For any given doublet $\mid a\pm \rangle$, one component interacts with $^3\Sigma_0{}^+$ and the other with $^3\Sigma_0{}^-$ states. A nonvanishing matrix element is equal to

$$(2)^{1/2}\langle[^3\Sigma 00][v'][JM_J0] \mid H_\Lambda \mid [^3\Pi 10][v][JM_J\ -1]\rangle, \tag{9.7}$$

where the pair of numbers following $^3\Sigma$ and $^3\Pi$ specify the values of Λ and

Σ. Using Eq. (9.5), we find that (9.7) becomes

$$-(2)^{-1/2}\langle[{}^3\Sigma 00][v'] \mid (\hbar^2/\mu r^2)(P_\xi - iP_\eta) \mid [{}^3\Pi 10][v]\rangle$$
$$\times \langle JM_J0 \mid J_\xi + iJ_\eta \mid JM_J\, -1\rangle$$
$$= \langle[{}^3\Sigma 0][v'] \mid (\hbar^2/\mu r^2) L_{0-1}^{(01)} \mid [{}^3\Pi 1][v]\rangle\{J(J+1)\}^{1/2}.$$

This produces a splitting in each J level of ${}^3\Pi_1$ proportional to $J(J+1)$.

The above analysis can be repeated for ${}^3\Pi_0$ and ${}^3\Pi_2$. However, neither ket $\mid {}^3\Pi 1\, -1\rangle$ or $\mid {}^3\Pi 11\rangle$ can be connected by **P** to $\langle {}^3\Sigma 00 \mid$. We can conclude that, to second order in perturbation theory, no Λ-doubling can be produced in the levels ${}^3\Pi_0$ and ${}^3\Pi_2$. This is not to say that no splittings are possible. Hebb [139] has pointed out that the matrix elements of the spin-spin interaction H_{ss} possess nonvanishing cross terms of the type

$$\langle[{}^3\Pi 1\, -1][v][JM_J0] \mid H_{ss} \mid [{}^3\Pi\, -1\, 1][v][JM_J0]\rangle, \qquad (9.8)$$

and these contribute to the energies of the two components of a doublet of ${}^3\Pi_0$ with opposite signs. The situation is very similar to that discussed in Section 7.4 for Hund's case (b). The main difference lies in the fact that the spins are referred to the molecular frame F' in Hund's case (a), with the result that H_{ss} is independent of the orientation of F' with respect to the laboratory. Integration over $(\phi\theta\gamma)$ in the matrix element (9.8) can be immediately carried out, and there is therefore no dependence on J or M_J.

To summarize, the Λ-doublings $\Delta E({}^3\Pi_{|\Omega|})$ are given, to second order, by

$$\Delta E({}^3\Pi_0) = C,$$
$$\Delta E({}^3\Pi_1) = C'J(J+1),$$
$$\Delta E({}^3\Pi_2) = 0, \qquad (9.9)$$

where C and C' are two constants. Further examples of Λ-doubling are discussed by Kovacs [90].

9.3 CORRECTIONS TO g_r

In Section 8.3 it was mentioned that the rotational parameters g_r for H_2, HD, and D_2 are found experimentally to be proportional to the inverses of the molecular reduced masses rather than to the right-hand side of Eq. (8.16). Following Ramsey [140], we look for second order effects which could resolve the difficulty. We begin by noticing that **L** has no nonvanishing matrix elements within the ground states ${}^1\Sigma$ of the three molecules in question. Hence the term in H_Z that involves g_r can be written as

$$H_{Zr} = -g_r\beta_N \mathbf{H}\cdot\mathbf{N}. \qquad (9.10)$$

We can readily see that the perturbation

$$\beta\mathbf{H}\cdot(\mathbf{L} + 2\mathbf{S}) - \hbar^2(\mathbf{N}\cdot\mathbf{L})/\mu r^2$$

could provide a cross term involving both $\mathbf{H}$ and $\mathbf{N}$ if, in second order, $\mathbf{L}r^{-2}$ and $\mathbf{L} + 2\mathbf{S}$ could connect $^1\Sigma$ to certain excited states. The contribution H_{Zr}' to H_{Zr} coming from this source can be formally written down without difficulty. Integrating over all coordinates except the Euler angles (which appear in $\mathbf{N}$), we get

$$H_{Zr}' = -\frac{2\beta\hbar^2}{\mu}\sum_n \frac{\mathbf{H}\cdot\langle 0 \mid \mathbf{L} + 2\mathbf{S} \mid n\rangle\langle n \mid r^{-2}\mathbf{L} \mid 0\rangle\cdot\mathbf{N}}{E_0 - E_n}, \tag{9.11}$$

where

$$|\,0\rangle = |\,[^1\Sigma][v]\rangle,$$
$$|\,n\rangle = |\,[\nu'\Pi\Lambda][v']\rangle \qquad (\Lambda = \pm 1), \tag{9.12}$$

and where E_0 and E_n are the corresponding energies (including both the electronic and vibrational contributions).

For the two matrix elements in Eq. (9.11) to be simultaneously nonvanishing, the components of $\mathbf{L} + 2\mathbf{S}$ and $\mathbf{L}$ must be either $L_+ + 2S_+$ and L_-, or $L_- + 2S_-$ and L_+. (We use the notation $T_+ = T_\xi + iT_\eta$, etc.) Both products of the associated matrix elements are of equal magnitude, as may be confirmed by using the reflection operator R of Section 6.9. We can therefore factor out

$$\tfrac{1}{2}(H_-N_+ + H_+N_-);$$

and since the eigenvalue of N_ζ is zero for $^1\Sigma$, we can add $H_\zeta N_\zeta$, thereby forming $\mathbf{H}\cdot\mathbf{N}$. In other words, we have

$$H_{Zr}' = -g_r'\beta_N\mathbf{H}\cdot\mathbf{N}, \tag{9.13}$$

where

$$g_r' = \frac{\beta\hbar^2}{\beta_N\mu}\sum_n \frac{\langle 0 \mid \mathbf{L} \mid n\rangle\cdot\langle n \mid r^{-2}\mathbf{L} \mid 0\rangle}{E_0 - E_n}. \tag{9.14}$$

We are able to form the scalar product of the two matrix elements because $L_\zeta\,|\,0\rangle = 0$. Equation (9.13) makes it obvious that the second order perturbations under consideration contribute to the experimentally measured values of g_r.

Implicit in Eqs. (9.10) and (9.14) is the choice of the center of mass C of the nuclei as the origin of coordinates. To exploit the identical electronic structure of H_2, HD, and D_2, the origin is transferred to M, the midpoint of the nuclear axis AB. This is done by taking

$$\boldsymbol{\varrho}_i = \boldsymbol{\varrho}_i' + \mathbf{a},$$

where $\mathbf{a} = \mathbf{r}_{MC}$. (Of course, $\mathbf{a} = 0$ for H_2 and D_2.) Thus

$$\mathbf{L} = \mathbf{L}' + \sum_t \mathbf{a} \times \mathbf{p}_t. \tag{9.15}$$

The two terms on the right-hand side of this equation possess opposite parities with respect to inversion in M. Thus no cross terms can arise when the replacement (9.15) is made for the two vectors $\mathbf{L}$ of Eq. (9.14). We get

$$g_r' = (\beta\hbar^2/\beta_N\mu)(\Xi + \Theta),$$

where

$$\Xi = \sum_n \frac{\langle 0 \mid \mathbf{L}' \mid n\rangle \cdot \langle n \mid r^{-2}\mathbf{L}' \mid 0\rangle}{E_0 - E_n},$$

$$\Theta = \sum_n \frac{\langle 0 \mid \sum \mathbf{a} \times \mathbf{p}_t \mid n\rangle \cdot \langle n \mid \sum \mathbf{a} \times \mathbf{p}_v r^{-2} \mid 0\rangle}{E_0 - E_n}.$$

The first sum, Ξ, possesses identical electronic matrix elements for the three molecules H_2, HD, and D_2. Only very slight variations due to different vibrational characteristics are expected. As for Θ, we can actually perform the sum over n by using Eq. (8.37), extended to a Hamiltonian that includes both the electronic and vibrational kinetic energy. We have

$$\langle 0 \mid \sum \mathbf{p}_t \mid n\rangle = (mi/\hbar^2)(E_0 - E_n)\langle 0 \mid \sum \boldsymbol{\varrho}_t' \mid n\rangle. \tag{9.16}$$

(A similar equation cannot be written for $\mathbf{p}_v r^{-2}$ because the vibrational kinetic energy does not commute with r^{-2}.) The closure relation [6]

$$\sum_n \mid n\rangle\langle n \mid = 1$$

allows us to write

$$\Theta = (mi/\hbar^2)\langle 0 \mid \sum_{tv} (\mathbf{a} \times \boldsymbol{\varrho}_t') \cdot (\mathbf{a} \times \mathbf{p}_v) r^{-2} \mid 0\rangle.$$

To conserve parity, we must have $v = t$. Furthermore, $\mathbf{a}$ is directed along the ζ axis, and this permits us to simplify the scalar product. We obtain

$$\Theta = (mi/\hbar^2)\langle 0 \mid (a^2 r^{-2}) \sum_t (\eta_t p_{t\eta} + \xi_t p_{t\xi}) \mid 0\rangle.$$

Partial integration yields, for any electron in a (real) σ orbital,

$$\left\langle \eta \frac{\partial}{\partial\eta} \right\rangle = \left\langle \xi \frac{\partial}{\partial\xi} \right\rangle = -\frac{1}{2}.$$

Since the sum over t comprises just two terms, we get

$$\Theta = (mi/\hbar^2)\langle 0 \mid a^2 r^{-2} \mid 0\rangle 4\left(\frac{1}{2} i\right) = -\frac{m}{2\hbar^2}\left(\frac{M_A - M_B}{M_A + M_B}\right)^2.$$

Quite remarkably, $\beta\hbar^2\Theta/\beta_N\mu$ combines with g_r to yield a term inversely proportional to μ:

$$M_P \frac{M_A{}^2 + M_B{}^2}{M_A M_B(M_A + M_B)} - \frac{\beta m}{2\beta_N\mu}\left(\frac{M_A - M_B}{M_A + M_B}\right)^2 = \frac{M_P(M_A + M_B)}{2M_A M_B} = \frac{M_P}{2\mu}.$$

Collecting our results together, we have

$$g_r + g_r' = \frac{M_P}{2\mu}\left(1 + \frac{2\hbar^2}{m}\,\Xi\right). \tag{9.17}$$

To the extent that Ξ is the same for H_2, HD, and D_2, the combined rotational parameter $g_r + g_r'$ depends inversely on μ, in good agreement with the experimental figures of Eqs. (8.19).

Second-order perturbation theory has not merely provided a correction to the first-order results: there has been an exact cancellation of those terms not proportional to μ^{-1}. This remarkable feature of the analysis recalls Van Vleck's demonstration that his formula for magnetic susceptibility be invariant to the choice of origin (see Problem 9.3). The overriding constraint that forces a μ^{-1} dependence appears to be the requirement that the energy of interaction of a constant magnetic field with a rigid rotating array of charges be invariant with respect to translations of the axis of rotation provided the sum of the charges be zero. This purely classical condition means that the interaction energy is proportional to the angular frequency $\tilde{\omega}$ of rotation; and, for a given total angular momentum, $\tilde{\omega}$ is in turn proportional to μ^{-1}. Of course, the four-particle system comprising H_2, HD, or D_2 is far from rigid; but at least the charge-density contours of the electronic cloud rotate with the two nuclei in a rigid fashion if vibrational effects are ignored. This appears to be enough to validate the classical model. However, one would expect to observe $g_r = -2.72$ for H_2 if the electronic charge cloud were totally rigid [119]. The electrons must evidently slide backwards over the rigid contours of their own charge distribution, since the observed figure of 0.88 is very close to the bare-nuclei value of 1.00.

9.4 DIAMAGNETIC SHIELDING

The presence of an electron cloud requires corrections not only to g_r but also to the purely nuclear terms $-g_X\beta_N\mathbf{H}\cdot\mathbf{I}_X$ in the Zeeman Hamiltonian H_Z of Eq. (8.15). Corrections of the latter kind are very much less significant than those for g_r, though the general approach is similar: we seek combinations of operators that can lead to products of $\mathbf{H}$ and $\mathbf{I}_X$. Before we look for second-order effects, however, we need to examine a source for diagonal corrections to H_Z. The effect of making the replacement

$$\mathbf{p}_t \to \mathbf{p}_t + (e/2\hbar c)(\mathbf{H} \times \mathbf{r}_t)$$

has been examined in Section 8.4 for spin–orbit contributions to the Hamiltonian; but it is clear that a precisely similar analysis can be carried out for the hyperfine-structure terms of Section 7.8. For the general diatomic molecule, the analog of Eq. (8.24) is

$$H_{\mathrm{ds}} = \tfrac{1}{2}\beta_N \alpha^2 a_0 \sum_{tX} g_X \mathbf{H}\cdot[\mathbf{I}_X(\mathbf{r}_{tX}\cdot\mathbf{r}_{tC}) - \mathbf{r}_{tX}(\mathbf{r}_{tC}\cdot\mathbf{I}_X)]r_{tX}^{-3}. \qquad (9.18)$$

Since H_{ds} is linear in $\mathbf{H}$ and $\mathbf{I}_X$, it can give rise to corrections to H_Z of the required kind. The atomic equivalent was first studied by Lamb [141], who interpreted the effect as diamagnetic shielding. The circulating electrons precess in the presence of the external magnetic field, and this leads to a reduction of the field at the atomic nucleus.

The analysis that led to Eqs. (8.29)–(8.30) now yields, for non-Π states,

$$H_{\mathrm{ds}} = \beta_N \sum_X [\sigma_{||}^X g_X H_\zeta I_{X\zeta} + \sigma_\perp^X g_X (H_\xi I_{X\xi} + H_\eta I_{X\eta})] \qquad (9.19)$$

$$= \sum_X \tfrac{1}{3}(\sigma_{||}^X + 2\sigma_\perp^X)\beta_N g_X \mathbf{H}\cdot\mathbf{I}_X$$
$$+ \sum_X \tfrac{1}{3}(2)^{1/2}(\sigma_\perp^X - \sigma_{||}^X)\beta_N g_X \mathbf{H}\cdot(\mathbf{D}^{(22)}\mathbf{I}_X^{(10)})_{\cdot 0}^{(12)}, \qquad (9.20)$$

where, in analogy to the expressions for $\Delta g_{||}$ and $\Delta g_\perp$,

$$\sigma_{||}^X = \tfrac{1}{2}\alpha^2 a_0 \langle \sum_t (\xi_t^2 + \eta_t^2) r_{tX}^{-3}\rangle,$$

$$\sigma_\perp^X = \tfrac{1}{2}\alpha^2 a_0 \langle \sum_t (\xi_t^2 + \zeta_{tX}\zeta_{tC}) r_{tX}^{-3}\rangle. \qquad (9.21)$$

For an atom comprising closed shells, we get

$$\sigma_{||}^X = \sigma_\perp^X = \sigma = \tfrac{1}{3}\alpha^2 a_0 \langle \sum_t r_t^{-1}\rangle,$$

in agreement with Lamb's result. Since r_t^{-1} is largest for the electrons closest to the nucleus, we might expect to get rather good values of $\sigma_{||}^X$ and $\sigma_\perp^X$ in the case of heavy molecules simply by taking the values of σ for the separate atomic constituents and ignoring departures from spherical symmetry.

In addition to these diagonal contributions to H_Z, however, terms involving $\mathbf{H}$ and $\mathbf{I}_X$ can also arise in second order perturbation theory from the combined action of the electronic Zeeman term $\beta\mathbf{H}\cdot(\mathbf{L} + 2\mathbf{S})$ and the hyperfine-structure interaction H_{hfs}. To illustrate how this works, it is convenient to simplify matters by supposing the molecule under investigation is in a $^1\Sigma$ state. The only effective part of the perturbation is

$$\beta \mathbf{H}\cdot\mathbf{L} + 2\beta\beta_N \sum_{tX} g_X r_{tX}^{-3}(\mathbf{r}_{tX} \times \mathbf{p}_t)\cdot\mathbf{I}_X.$$

The contribution H_{ds}' that should be added to H_{ds} is

$$4\beta^2\beta_N \sum_n \frac{\mathbf{H}\cdot\langle 0 \mid \mathbf{L} \mid n\rangle\langle n \mid \sum g_X r_{tX}^{-3}(\mathbf{r}_{tX} \times \mathbf{p}_t) \mid 0\rangle\cdot\mathbf{I}_X}{E_0 - E_n}, \tag{9.22}$$

where the kets $\mid 0\rangle$ and $\mid n\rangle$ are defined as in Eqs. (9.12). Both states possess the same parity, for otherwise the matrix element of **L** would vanish; and this allows us to make the replacements

$$\mathbf{r}_{tX} \times \mathbf{p}_t \rightarrow \mathbf{r}_{tC} \times \mathbf{p}_t = \boldsymbol{\lambda}_t.$$

There is no $H_\zeta I_{X\zeta}$ term in the expression (9.22), since $\langle 0 \mid L_\zeta = 0$. We find that the effect of the perturbation can be represented by replacing $\sigma_\perp^X$ everywhere by

$$\sigma_\perp^X + 2\beta^2 \sum_n \frac{\langle 0 \mid \mathbf{L} \mid n\rangle\cdot\langle n \mid \sum r_{tX}^{-3}\boldsymbol{\lambda}_t \mid 0\rangle}{E_0 - E_n}.$$

No correction to $\sigma_{||}^X$ is required. An equivalent result has been given by Ramsey [119].

The first sum over X on the right-hand side of Eq. (9.20) combines with the $\mathbf{H}\cdot\mathbf{I}_X$ terms in H_Z. It can thus be taken into account by using the original Hamiltonian H_Z and simply introducing screened g_X values. The second sum over X, however, involves the tensor $\mathbf{D}^{(22)}$. Being a function of the Euler angles, it gives rise to a dependence on the angular momentum J. (For a $^1\Sigma$ state, $J = N$.) By comparing reduced matrix elements for a rotational state of given J, we can rapidly show that

$$\mathbf{D}_{\cdot 0}^{(22)} \equiv -\frac{(30)^{1/2}}{(2J-1)(2J+3)}(\mathbf{JJ})^{(2)}.$$

If we suppose that the magnetic field **H** is applied along the z direction, so that the molecular states produced by the Zeeman Hamiltonian H_Z can be written as

$$\mid JM_J, I_AM_{IA}, I_BM_{IB}\rangle,$$

then only the diagonal matrix elements of the small perturbative Hamiltonian H_{ds} are required. In this case, the only part of

$$\{(\mathbf{JJ})^{(2)}\mathbf{I}_X^{(1)}\}^{(1)}$$

that can contribute is

$$(20, 10 \mid 10)\,(\mathbf{JJ})_0{}^{(2)} I_{X0}{}^{(1)},$$

and we can write, for this special case,

$$H_{\mathrm{ds}} = \sum_X \sigma^X \beta_N g_X H I_{X0}{}^{(1)},$$

where

$$\sigma^X = \frac{1}{3}(\sigma_{||}{}^X + 2\sigma_{\perp}{}^X) + \frac{2}{3(2J-1)(2J+3)}(\sigma_{\perp}{}^X - \sigma_{||}{}^X)[3M_J{}^2 - J(J+1)].$$

This is consistent with Ramsey's original expression [119]. Some calculations of $\sigma_{\perp}{}^X - \sigma_{||}{}^X$, as well as a preliminary experimental figure, have been reported [142].

9.5 HOMOMORPHIC PERTURBATIONS

One of the characteristic features of perturbation analyses is that second or higher order processes often give rise to energy-level displacements that could be reproduced by making adjustments to the parameters that describe lower order effects. For example, it has been seen in Section 9.3 that the cross term coming in second order from $\mathbf{H}\cdot\mathbf{L}$ and $\mathbf{L}\cdot\mathbf{N}$ contributes to the first order term $\mathbf{H}\cdot\mathbf{N}$ in the Zeeman Hamiltonian for a $^1\Sigma$ state. Results of this kind are a common feature not only of molecular spectroscopy but of atomic and nuclear spectroscopy too. Although the methods of Sections 9.3 and 9.4 are usually adequate to throw the second order contributions into an acceptable form, it is sometimes useful to regard the energy displacements as being produced by an operator Z, defined by

$$Z = VQV, \tag{9.23}$$

where

$$Q = \sum_{n\neq 0} \frac{|n\rangle\langle n|}{E_0 - E_n}. \tag{9.24}$$

The operator Z, when sandwiched between the bra $\langle 0|$ and the ket $|0\rangle$ of the state under investigation, gives the second order energy corrections due to a perturbation V.

As any other operator, Z can be examined for its tensorial characteristics.

If we are working in case (b) coupling, the sum over n in Q contains many internal sums of the type

$$\sum_{M_S} | SM_S\rangle\langle SM_S | .$$

This expression commutes with $\mathbf{S}$, as we may rapidly verify. Provided all the M_S components of a spin S share the same energy E_n, we can regard Q as a scalar with respect to spin. For similar reasons, Q can be considered to be a scalar with respect to $\mathbf{N}^{(10)}$ if all the M components of N correspond to the same E_n. Since these conditions are broken only by the fine structure of a rotational level—which is almost always extremely small compared to $E_0 - E_n$—the scalar character of Q with respect to $\mathbf{S}$ and $\mathbf{N}^{(10)}$ is assured. We would not, on the other hand, expect it to be a scalar with respect to the operator $\mathbf{G}^{(01)}$ of Section 6.4, since the eigenvalues Γ of G_ζ are limited to the value zero. Of course, Q can also be examined with respect to other operators, such as the reflection operation R. In this case, it is clear that we can assume Q commutes with R, because the two components of a Λ-doubled level are separated by an energy extremely small compared to $E_0 - E_n$.

Knowing the characteristics of Q, we can construct the product VQV and hence determine the properties of Z. The matrix elements of Z often assume a comparatively simple form when $|0\rangle$ is a Σ state. This is apparent from the analysis of Section 9.3, where the $^1\Sigma$ state of H_2 is studied. Another classic example is the ground state $^3\Sigma$ of O_2. Just as with the analysis of $^3\Pi$ described in Section 7.4, every rotational level is a triplet. However, the spin–orbit interaction cannot now contribute in first order because $\Lambda = 0$ for a Σ state. The triplet fine structure is thus composed of a spin–spin part and a small contribution from the spin–rotation interaction. Equations (7.16) and (7.36) allow us to relate the diagonal matrix elements E_{N1J} of the fine-structure Hamiltonian $H_{ss} + H_{sr}$ through the equations

$$E_{N1N} - E_{N1\,N+1} = 2\lambda\{(N+1)/(2N+3)\} - \mu(N+1),$$

$$E_{N1N} - E_{N1\,N-1} = 2\lambda\{N/(2N-1)\} + \mu N, \tag{9.25}$$

where λ and μ are parameters corresponding to H_{ss} and H_{sr} respectively. Equations (9.25), which were first obtained by Kramers [143], give the first-order separations of the components of a triplet.

If, now, we turn to second-order contributions to E_{N1J}, we see at once that the spin-orbit interaction H_{so} can play a role by admixing other electronic states into $^3\Sigma$. Thus Z is a sum of operators of the type

$$f = \mathbf{u}_{vt}\cdot\mathbf{s}_t Q \mathbf{u}_{wp}\cdot\mathbf{s}_p,$$

where the subscripts refer to the electrons. The operator f, as it stands, cannot be characterized by a single spin rank. Since Q is a spin scalar, a recoupling can only produce operators whose spin ranks are 0, 1, and 2. That is, we can write

$$f = f_0 + f_1 + f_2,$$

where f_κ behaves as a tensor of rank κ with respect to $\mathbf{S}$. The operator f_κ is also a tensor of rank κ with respect to $\mathbf{N}^{(10)}$, since f is a scalar with respect to $\mathbf{J}$. Each part f_κ can now be examined in turn.

The scalar, f_0, is easy to treat, because it displaces all three components of an NS triplet equally. As we are only interested in splittings, this can be discarded.

Passing to f_1, we recall (from Section 9.1) that the ζ component of $\mathbf{u}_{vt}$ changes sign under the reflection R. In greater detail,

$$R(u_{0q}{}^{(01)})_{vt}R^{-1} = (-1)^{q+1}(u_{0-q}{}^{(01)})_{vt}.$$

From this we can deduce that

$$R(\mathbf{u}_{vt}{}^{(01)}\mathbf{u}_{wp}{}^{(01)})_{0q}{}^{(0k)}R^{-1} = (-1)^{q+k}(\mathbf{u}_{vt}{}^{(01)}\mathbf{u}_{wp}{}^{(01)})_{0-q}{}^{(0k)}.$$

For Σ states, only the component for which $q = 0$ can contribute to a matrix element. Thus

$$\begin{aligned}\langle\Sigma \mid (\mathbf{u}_{vt}{}^{(10)}\mathbf{u}_{wp}{}^{(10)})^{(10)} \mid \Sigma\rangle &= [\mathbf{D}^{(11)}\langle\Sigma \mid (\mathbf{u}_{vt}{}^{(01)}\mathbf{u}_{wp}{}^{(01)})^{(01)} \mid \Sigma\rangle]^{(10)} \\ &= 0.\end{aligned}$$

The two $\mathbf{u}$ tensors in f are separated by Q; but since Q is a scalar with respect to R, its intervention cannot affect the conclusion that f_1 has null matrix elements.

We are thus left with f_2. Being of rank 2 with respect to both S and N, its matrix elements must depend on J in the same way as those of H_{ss}. Their dependence on N must also be identical, since the Euler angles enter in the same way—namely, through the tensor $\mathbf{D}^{(22)}$—for f_2 as for H_{ss}. Hence the second order process involving H_{so} twice can only contribute to the triplet splittings through an adjustment to λ. It is convenient to write

$$\lambda = \lambda_{ss} + \lambda_{so}$$

to separate the two parts of the parameter λ. The first detailed calculations of λ were carried out by Tinkham and Strandberg [130], who estimated that $\lambda_{ss} = 1.2$ cm^{-1}. More recent calculations by Pritchard *et al.* [144] set λ_{ss} at 0.77 cm^{-1}. These writers also established that this figure may be lowered still further by configuration interaction. Since the experimental value for λ is 1.98 cm^{-1}, a substantial contribution must come from λ_{so}. The principal source appears to be $^1\Sigma_g{}^+$, which supplies 1.25 cm^{-1} to λ_{so}, according to the calculations of Hall [145]. Presumably the effect of other excited states is small.

A similar analysis reveals that second order perturbation theory involving H_{so} and the $\mathbf{N}\cdot\mathbf{L}$ term in H_r provides a contribution to E_{N1J} that can be absorbed by a change in μ. The fact that the form of Eqs. (9.25) remains valid when certain second order effects are taken into account was first demonstrated by Hebb [139].

9.6 CENTRIFUGAL CORRECTIONS

The discussion of the rotating oscillator in Section 6.7 indicates that centrifugal and anharmonic effects can be taken into account in zeroth order by solving Eq. (6.36) with the appropriate potential function $W(r)$ given in Eq. (6.37). However, the basic states that we have been using (i.e., $|a\pm\rangle$ or $|b\pm\rangle$) contain parts such as

$$|[\nu\Lambda S\Sigma][v][JM_J -\Omega]\rangle \qquad (9.26)$$

that are simple products of a rotational eigenfunction $|JM_J -\Omega\rangle$ and a perfectly harmonic vibrational eigenfunction $|v\rangle$. For a consistent theory, it must be possible to obtain the centrifugal and anharmonic effects by applying perturbation theory to the alternative basis provided by the states $|a\pm\rangle$ or $|b\pm\rangle$. Hougen [146] has shown how this approach accounts for the centrifugal corrections to the rotational energy levels, and it seems worthwhile to give an abbreviated form of his analysis here. We shall concentrate our attention on a possible source for a term proportional to $J^2(J+1)^2$.

The rotational Hamiltonian H_r' (for Hund's case (a)) is given in Eq. (6.51) as

$$H_r' = (\hbar^2/2\mu r^2)(\mathbf{J}-\mathbf{P})^2.$$

In preparation for the calculation of matrix elements of H_r' between different vibrational states, we write

$$r^{-2} = (r_e + x)^{-2} = r_e^{-2} - 2xr_e^{-3} + 3x^2r_e^{-4} - \cdots,$$

where the displacement x (assumed small) is simply $r - r_e$. So, to first order,

$$H_r' = B(1 - 2xr_e^{-1})(\mathbf{J}-\mathbf{P})^2,$$

where $B = \hbar^2/2\mu r_e^2$. If attention is limited to the vibrational levels coming from a single electronic state, only the term

$$-2Bxr_e^{-1}(\mathbf{J}-\mathbf{P})^2 \qquad (9.27)$$

can give contributions off-diagonal with respect to v. Our preoccupation in the term in $\mathbf{J}^4$ leads us to retain only the $\mathbf{J}^2$ term in $(\mathbf{J} - \mathbf{P})^2$; and since the matrix elements of $\mathbf{J}^2$ are independent of x, the only nonvanishing matrix elements we need are

$$\langle v \mid x \mid v+1 \rangle = \{(v+1)/2\kappa\}^{1/2},$$
$$\langle v \mid x \mid v-1 \rangle = (v/2\kappa)^{1/2},$$

where $\kappa = \mu\omega/\hbar$, as in Section 6.6. So

$$\sum_{v'} \langle v \mid x \mid v' \rangle \langle v' \mid x \mid v \rangle / (E_v - E_{v'}) = -(2\kappa\hbar\omega)^{-1}$$
$$= -(2\mu\omega^2)^{-1}.$$

The actual perturbation is not simply x, but rather the expression (9.27). The evaluation of $\mathbf{J}^2$ between two pairs of rotational states gives rise to a term

$$-CJ^2(J+1)^2$$

in the expression for the rotational energy, where

$$C = 4B^2 r_e^{-2}(2\mu\omega^2)^{-1} = 4B^3/\hbar^2\omega^2.$$

It has already been mentioned (in Problem 6.4) that this equation (adjusted by multiplying B and C by $\hbar$) obtains for both the Morse potential and the rather artificial potential $W(r)$ of Eq. (6.42). We can now see that it is essentially rotational in origin and thus independent of the particular form of $W(r)$.

9.7 SPECTROSCOPIC PERTURBATIONS

Since the array of rotational levels associated with every electronic state has no upper bound (other than that determined by the condition for dissociation of the molecule), overlapping sequences of rotational levels are a commonplace. It sometimes happens that a pair of levels lie sufficiently close to each other to be perceptibly perturbed by their mutual interaction, thereby producing displacements of the components of a band as well as intensity anomalies. The strength of the interaction is determined by the off-diagonal matrix elements of the total Hamiltonian, H. If we neglect hyperfine effects, H commutes with $\mathbf{J}$; thus only levels with the same J can interact. Further selection rules can be obtained from the various parts that contribute to H. If all terms involving spin are sufficiently small to be set aside, then H commutes with L_ζ and $\mathbf{S}$. The interacting states must now satisfy the selection rules $\Delta\Lambda = 0$, $\Delta S = 0$. The inclusion of spin terms that transform as vectors in the spin space yield the less stringent rules

$$\Delta\Lambda = 0, \pm 1, \qquad \Delta S = 0, \pm 1.$$

The spin-spin interaction H_{ss}, which is a tensor of rank 2 in the spin space, would extend the range of $\Delta\Lambda$ and ΔS to ± 2; however, the effect of H_{ss} is usually sufficiently small to be neglected. Perturbations corresponding to $\Delta\Lambda = 0$ are called *homogeneous*; those corresponding to $\Delta\Lambda = \pm 1$ are called *heterogeneous*. An extensive survey of both kinds of perturbations has been made by Kovacs [90]. We limit ourselves to a few general remarks and an example.

An important difference between homogeneous and heterogeneous perturbations is that the former contain parts—such as the vibrational Hamiltonian H_v—whose matrix elements are independent of J. These frequently outweigh the effects of the terms arising from spin interactions, which exhibit a dependence on J. In contrast to this, the spin interactions constitute the sole source of heterogeneous perturbations, which are therefore J-dependent. To calculate the actual energies of the levels of two mutually perturbing series (distinguished by the symbols α and β), we have only to diagonalize the matrices

$$\begin{pmatrix} \langle \alpha J \mid H \mid \alpha J \rangle & \langle \alpha J \mid H \mid \beta J \rangle \\ \langle \beta J \mid H \mid \alpha J \rangle & \langle \beta J \mid H \mid \beta J \rangle \end{pmatrix}$$

for the values of J we are interested in. The entries on the diagonal follow the usual formulae: for example,

$$\langle \alpha J \mid H \mid \alpha J \rangle = A_\alpha + B_\alpha J(J+1) - D_\alpha J^2(J+1)^2.$$

Attention therefore centers on the off-diagonal matrix elements.

As an example, we pick

$$\alpha \equiv [\nu\,^2\Sigma][v], \qquad \beta \equiv [\nu'\,^2\Pi_{|\Omega|}][v'].$$

The perturbation is heterogeneous with two principal sources: the spin–orbit interaction H_{so} and the rotational term H_r'. As the notation indicates, β approximates to Hund's case (a), for which the states $| a\pm \rangle$ of Eq. (6.63) are appropriate. The $^2\Sigma$ state, however, is invariably represented best by Hund's case (b) coupling. To conserve parity, only one of the two states $| a\pm \rangle$ can interact with a particular level N of $^2\Sigma$; and, for this one,

$$\begin{aligned}\langle \alpha J \mid H \mid \beta J \rangle \\ &= (2)^{1/2}\langle [^2\Sigma][v][N0]SJM_J \mid H_{so} \\ &\qquad + H_r' \mid [^2\Pi, 1, \Omega - 1][v'][JM_J\, -\Omega] \rangle.\end{aligned}$$

We transform the $^2\Sigma$ state to a sum over states in case (a) coupling by means of the equation

$$\langle[^2\Sigma][v][N0]SJM_J| = \sum_{\Omega'} (N0\,|\,S\Omega', J\,-\Omega')\,\langle[^2\Sigma, 0, \Omega'][v][JM_J\,-\Omega']|,$$

which follows from Problem 9.1. The term $\mathbf{u}_{vt}\cdot\mathbf{s}_t$, which has already been used in Section 9.1 as the prototype of a component of H_{so}, is diagonal with respect to Ω (since $\mathbf{u}^{(01)}\cdot\mathbf{s}^{(01)}$ commutes with $\mathbf{J}^{(01)}$); however, the $\mathbf{J}\cdot\mathbf{L}$ component of H_r' satisfies $\Delta\Omega = \pm 1$, since only $L_\xi \pm iL_\eta$ is effective in coupling $^2\Sigma$ to $^2\Pi$. Putting $\mathbf{s}_t \equiv \kappa\mathbf{S}$ (as we did in Section 8.4), we find

$$\begin{aligned}\langle\alpha J\,|\,H\,|\,\beta J\rangle = \frac{1}{2}(\Theta + \Phi)(-1)^{J-1/2}\{(2N+1)(2\Omega+1)(3-2\Omega)\}^{1/2}\\ \times\begin{pmatrix} N & \frac{1}{2} & J \\ 0 & \Omega & -\Omega\end{pmatrix}\\ + \Phi(-1)^{J-1/2}\{(2N+1)(J+1-\Omega)(J+\Omega)\}^{1/2}\\ \times\begin{pmatrix} N & \frac{1}{2} & J \\ 0 & \Omega-1 & -\Omega+1\end{pmatrix},\end{aligned} \tag{9.28}$$

where

$$\Theta = \kappa\langle[^2\Sigma 0][v]\,|\,\textstyle\sum(u_{0-1}{}^{(01)})_{vt}\,|\,[^2\Pi 1][v']\rangle,$$

$$\Phi = \langle[^2\Sigma 0][v]\,|\,(\hbar^2/\mu r^2)L_{0-1}{}^{(01)}\,|\,[^2\Pi 1][v']\rangle.$$

The Φ term in the combination $(\Theta + \Phi)$ that occurs on the right-hand side of Eq. (9.28) derives from the $\mathbf{S}\cdot\mathbf{L}$ term in H_r'. Explicit forms of $\langle\alpha J\,|\,H\,|\,\beta J\rangle$ for the various possible values of N and $|\,\Omega\,|$ (namely $J \pm \frac{1}{2}$ and $1 \pm \frac{1}{2}$ respectively) are given by Kovacs [90]. (For the example in hand, his ξ and η are related to our Θ and Φ by the equations $\Theta = \xi\sqrt{2}$, $\Phi = 2\eta\sqrt{2}$.) Kovacs also describes an interesting case of $^2\Sigma - {}^2\Pi$ perturbations for the vibrational states $v = 11$ and $v' = 4$ of the CN molecule.

9.8 FORBIDDEN TRANSITIONS

In molecular spectroscopy, it is common to find transitions that, at first sight, violate the selection rules. The explanation lies in the fact that states are labeled by their principal components, and it is the small admixtures of other states that permit the transition to take place. We have already mentioned the transition $^1\Sigma \rightarrow {}^3\Pi$ in I_2 (see Section 8.9). To investigate the procedure for calculating relative intensities, we choose the transition

${}^1\Delta_g \rightarrow {}^3\Sigma_g$ of O_2, which is important for atmospheric physics [147]. Since $\Delta S = 1$, the transition is forbidden for electric-dipole, magnetic-dipole, and electric-quadrupole radiation. However, the gap between ${}^1\Delta_g$ and ${}^3\Sigma_g$ can be bridged by the magnetic-dipole operator $\mathbf{M}$ ($\equiv \mathbf{L} + 2\mathbf{S}$) if the spin–orbit interaction H_{so} acts in concert. Symbolically, two linkages are possible:

$$\langle {}^1\Delta_g \mid \mathbf{M} \mid {}^1\Pi_g \rangle \langle {}^1\Pi_g \mid H_{so} \mid {}^3\Sigma_g \rangle,$$
$$\langle {}^1\Delta_g \mid H_{so} \mid {}^3\Pi_g \rangle \langle {}^3\Pi_g \mid \mathbf{M} \mid {}^3\Sigma_g \rangle. \tag{9.29}$$

Of course, the relative intensities of the lines making up a band are determined by the dependence of these products on quantum numbers that have not been explicitly included here. In Hund's case (b) coupling, these are J and N for ${}^3\Sigma$, and J' ($=N'$) for ${}^1\Delta$. The corresponding quantum numbers (J'' and N'', say) for the intermediate states (${}^1\Pi$ and ${}^3\Pi$) can be removed from the analysis by performing a limited closure over just the rotational eigenfunctions. This is possible because the associated energy denominators are determined almost entirely by the differences in the electronic energies, and are quite insensitive to the slight changes in the rotational energies.

Having eliminated N'' and J'', we can bring together $\mathbf{M}$ and H_{so} provided our interest lies only in the dependence of the products (9.29) on N, J, N', and J'. From the combination

$$M_q \sum \mathbf{u}_{vt} \cdot \mathbf{s}_t,$$

the only part that can directly connect a Δ state to a Σ state is

$$(\mathbf{W}^{(2)}\mathbf{T}^{(1)})_q{}^{(1)},$$

where $\mathbf{W}^{(2)}$ acts on the orbit and $\mathbf{T}^{(1)}$ on the spin. Thus, to within a constant (independent of N, J, N', and J'), we can write, for the line strength $\mathfrak{L}_{md}$ of the magnetic-dipole radiation,

$$\mathfrak{L}_{md} = | (\Delta(N'S')J' |\!: (\mathbf{W}^{(2)}\mathbf{T}^{(1)})^{(1)} :\!| \Sigma(NS)J) |^2,$$

in which $S' = 0$ and $S = 1$. Using Eq. (1.62), we get

$$\mathfrak{L}_{md} = 3[J', J] \begin{Bmatrix} N' & N & 2 \\ 0 & 1 & 1 \\ J' & J & 1 \end{Bmatrix}^2 (\Delta N' |\!: W^{(20)} :\!| \Sigma N)^2 (S' \| T^{(1)} \| S)^2$$

$$= (2J+1) \begin{Bmatrix} J & 1 & J' \\ 2 & N & 1 \end{Bmatrix}^2 \delta(N', J') (S' \| T^{(1)} \| S)^2$$

$$\times (\Delta N' |\!: (\mathbf{D}^{(22)}\mathbf{W}^{(02)})^{(20)} :\!| \Sigma N)^2.$$

TABLE 9.1 Line Strengths for ${}^1\Delta(N'J') \rightarrow {}^3\Sigma(NJ)$

Branch	N' ($=J'$)	N	$\mathfrak{L}_{md}'$
${}^OP(J)$	$J-1$	$J+1$	$\dfrac{(J-1)(J-2)}{4(2J+1)}$
${}^PP(J)$	$J-1$	J	$\dfrac{(J-1)(J-2)}{4J}$
${}^QP(J)$	$J-1$	$J-1$	$\dfrac{(J+1)(J-1)(J-2)}{4J(2J+1)}$
${}^PQ(J)$	J	$J+1$	$\dfrac{(J-1)(J+2)}{4(J+1)}$
${}^QQ(J)$	J	J	$\dfrac{(J-1)(J+2)(2J+1)}{4J(J+1)}$
${}^RQ(J)$	J	$J-1$	$\dfrac{(J-1)(J+2)}{4J}$
${}^QR(J)$	$J+1$	$J+1$	$\dfrac{J(J+2)(J+3)}{4(J+1)(2J+1)}$
${}^RR(J)$	$J+1$	J	$\dfrac{(J+2)(J+3)}{4(J+1)}$
${}^SR(J)$	$J+1$	$J-1$	$\dfrac{(J+2)(J+3)}{4(2J+1)}$

Only the components $\mathbf{D}_{\cdot\pm 2}^{(22)}$ of $\mathbf{D}^{(22)}$ can contribute. Dropping factors that do not depend on N, J, N', and J', we arrive at the result $\mathfrak{L}_{md} \propto \mathfrak{L}_{md}'$, where

$$\mathfrak{L}_{md}' = 5[J, N, N']\delta(N', J') \begin{Bmatrix} J & 1 & J' \\ 2 & N & 1 \end{Bmatrix}^2 \begin{pmatrix} N' & 2 & N \\ 2 & -2 & 0 \end{pmatrix}^2. \qquad (9.30)$$

The factor of 5 is included to ensure that Eq. (9.30) gives results identical to the special cases listed by Kovacs [90]. A slightly different normalization was used by Van Vleck in his original calculation [148]. Explicit expressions for the line strengths $\mathfrak{L}_{md}'$ given in Eq. (9.30) are set out in Table 9.1 with the appropriate labels for the branches. Values of ΔN are indicated by lettered prefixes to the usual P, Q, and R designations. It is interesting to notice that the tensorial characteristics of $(\mathbf{W}^{(2)}\mathbf{T}^{(1)})^{(1)}$ completely determine the relative line strengths. This property would permit us to distinguish the magnetic-dipole transitions from others that might contribute to the intensities. For example, we could imagine quadrupole radiation induced by H_{so}; this would provide two sources corresponding to two coupled tensors of the types

$$(\mathbf{W}^{(2)}\mathbf{T}^{(1)})^{(2)}, \qquad (\mathbf{W}^{(3)}\mathbf{T}^{(1)})^{(2)},$$

and more branches would occur.

PROBLEMS

9.1 Show that a state corresponding to Hund's case (a) can be expanded in states corresponding to case (b) by writing

$$\begin{aligned} &|[\nu\Lambda S\Sigma][v][JM\ -\Omega]\rangle \\ &\quad = \sum_N (N\ -\Lambda \mid S\Sigma, J\ -\Omega) \mid [\nu\Lambda][v][N\ -\Lambda]SJM\rangle. \end{aligned}$$

Prove that the secular determinants on Section 9.1 yield eigenfunctions that tend towards the case (b) description as $J \to \infty$.

9.2 Show that, to second order in perturbation theory, and in the limit of Hund's case (a), the rotational levels of $^2\Pi_{3/2}$ exhibit no Λ-doubling, while those of $^2\Pi_{1/2}$ are split by an amount proportional to $2J + 1$ (see Van Vleck [149] and Kovacs [90]).

9.3 Show that the $\mathbf{A}^2$ term in the Zeeman Hamiltonian for a diatomic molecule can be expressed as

$$H_{ms} = \sum_t (e^2/8mc^2)[\tfrac{2}{3}H^2 r_{tC}^2 - (\mathbf{HH})^{(2)} \cdot (\mathbf{r}_{tC}\mathbf{r}_{tC})^{(2)}].$$

The molecular magnetic susceptibility χ is defined in terms of an average energy change δW through the equation

$$\delta W = -\chi \mathbf{H} \cdot \delta \mathbf{H}.$$

Assuming that the components of any rotational level of a $^1\Sigma$ state have an equal probability of being occupied, show that the magnetic susceptibility Hamiltonian H_{ms} leads to

$$\chi = -(e^2/6mc^2)\langle {}^1\Sigma \mid \sum_t \rho_t^2 \mid {}^1\Sigma\rangle.$$

Prove that the interaction $\beta\mathbf{H}\cdot(\mathbf{L} + 2\mathbf{S})$ produces a second order correction χ' to χ given by

$$\chi' = -\frac{2}{3}\beta^2 \sum_\nu \frac{|\langle {}^1\Sigma | \mathbf{L} | \nu {}^1\Pi \rangle|^2}{E({}^1\Sigma) - E(\nu {}^1\Pi)},$$

and verify that $\chi + \chi'$ is independent of the choice of origin (see Van Vleck [150]).

9.4 If matrix elements of H_{ss} off-diagonal with respect to N are taken into account, show that Eqs. (9.25) become, for a given rotational level of $O_2\ {}^3\Sigma^-$,

$$\begin{aligned} E_{N1N} - E_{N1\,N+1} &= -B(2N+3) + \lambda + \tfrac{1}{2}\mu \\ &\quad + [(2N+3)^2(B - \tfrac{1}{2}\mu)^2 + \lambda^2 - 2\lambda(B - \tfrac{1}{2}\mu)]^{1/2}, \\ E_{N1N} - E_{N1\,N-1} &= B(2N-1) + \lambda + \tfrac{1}{2}\mu \\ &\quad - [(2N-1)^2(B - \tfrac{1}{2}\mu)^2 + \lambda^2 - 2\lambda(B - \tfrac{1}{2}\mu)]^{1/2}, \end{aligned}$$

where the rotational parameter B enters in the usual way through the expression $BJ(J+1)$ for the rotational energy for the level J (see Miller and Townes [151]). Show that the effect of the spin–orbit interaction H_{so} taken to second order can be absorbed by adjusting λ.

9.5 Show that the spin–spin interaction H_{ss} can be represented by the effective operator

$$\tfrac{2}{3}\lambda(3S_\zeta^2 - \mathbf{S}^2)$$

within the ground state ${}^3\Sigma^-$ of O_2, where λ has the same significance as in Eqs. (9.25) (see Tinkham and Strandberg [130]).

9.6 A diatomic molecule is subjected to a constant external electric field $\mathbf{E}$. Prove that, in addition to the linear Stark effect (described in Section 8.2), the interactions between neighboring rotational levels of a given vibronic state in Hund's case (a) coupling produce the quadratic displacements

$$\frac{E^2\mu_e^2}{2B}\left[\frac{(J^2 + J - 3M_J^2)(J^2 + J - 3\Omega^2)}{J^2(J+1)^2(2J-1)(2J+3)} - \frac{\Omega^2 M_J^2}{J^3(J+1)^3}\right],$$

where μ_e is defined as in Eq. (8.11) and B is the rotational constant (as in Problem 9.4) (see Townes and Schawlow [152]).

9.7 Use Eq. (9.30) to prove that

$$\sum_{N'J'N} \mathfrak{L}_{\text{md}}' = 2J + 1,$$

and check this result from Table 9.1.

Appendix

1

Algebraic Expressions for Certain n-j Symbols

An explicit formula for the 3-j symbol can be constructed by combining Eq. (1.31) with the expression for a CG coefficient given in Problem 1.3. For the 6-j symbol, we have only to refer to Eq. (1.64) to get an explicit result. When one of the angular momentum quantum numbers in an n-j symbol is small, the general expression simplifies appreciably. The list of 3-j and 6-j symbols that follows contains all the possibilities that can occur when one of these quantum numbers is 0, $\frac{1}{2}$, or 1. A more extensive tabulation has been provided by Edmonds [1].

$$\begin{pmatrix} J & J & 0 \\ M & -M & 0 \end{pmatrix} = (-1)^{J-M} \frac{1}{(2J+1)^{1/2}},$$

$$\begin{pmatrix} J+\frac{1}{2} & J & \frac{1}{2} \\ M-\frac{1}{2} & -M & \frac{1}{2} \end{pmatrix} = (-1)^{J-M} \left[\frac{J-M+1}{(2J+1)(2J+2)}\right]^{1/2},$$

$$\begin{pmatrix} J+1 & J & 1 \\ M-1 & -M & 1 \end{pmatrix} = (-1)^{J-M}\left[\frac{(J-M+1)(J-M+2)}{(2J+1)(2J+2)(2J+3)}\right]^{1/2},$$

$$\begin{pmatrix} J+1 & J & 1 \\ M & -M & 0 \end{pmatrix} = (-1)^{J-M-1}\left[\frac{(J-M+1)(J+M+1)}{(J+1)(2J+1)(2J+3)}\right]^{1/2},$$

$$\begin{pmatrix} J & J & 1 \\ M & -M-1 & 1 \end{pmatrix} = (-1)^{J-M}\left[\frac{(J-M)(J+M+1)}{2J(J+1)(2J+1)}\right]^{1/2},$$

$$\begin{pmatrix} J & J & 1 \\ M & -M & 0 \end{pmatrix} = (-1)^{J-M}\frac{M}{[J(J+1)(2J+1)]^{1/2}}.$$

Other examples of 3-j symbols can be derived from the ones above by making use of the symmetry relations

$$\begin{pmatrix} A & B & C \\ a & b & c \end{pmatrix} = (-1)^{A+B+C}\begin{pmatrix} A & B & C \\ -a & -b & -c \end{pmatrix}$$

$$= (-1)^{A+B+C}\begin{pmatrix} A & C & B \\ a & c & b \end{pmatrix} = \begin{pmatrix} B & C & A \\ b & c & a \end{pmatrix},$$

etc. When $a = b = c = 0$, the 3-j symbol vanishes unless $A + B + C$ (which we denote by J) is even: in that case

$$\begin{pmatrix} A & B & C \\ 0 & 0 & 0 \end{pmatrix} = (-1)^{\frac{1}{2}J}\left[\frac{(J-2A)!(J-2B)!(J-2C)!}{(J+1)!}\right]^{1/2}$$

$$\times \frac{(\frac{1}{2}J)!}{(\frac{1}{2}J-A)!(\frac{1}{2}J-B)!(\frac{1}{2}J-C)!}.$$

In listing the 6-j symbols with one argument equal to 0, $\frac{1}{2}$, or 1, it is convenient to make the abbreviation $J = A + B + C$ again:

$$\begin{Bmatrix} A & B & C \\ 0 & C & B \end{Bmatrix} = (-1)^{J}\left[\frac{1}{(2B+1)(2C+1)}\right]^{1/2},$$

$$\begin{Bmatrix} A & B & C \\ \frac{1}{2} & C-\frac{1}{2} & B+\frac{1}{2} \end{Bmatrix} = (-1)^{J}\left[\frac{(J-2B)(J-2C+1)}{(2B+1)(2B+2)2C(2C+1)}\right]^{1/2},$$

$$\begin{Bmatrix} A & B & C \\ \frac{1}{2} & C-\frac{1}{2} & B-\frac{1}{2} \end{Bmatrix} = (-1)^{J}\left[\frac{(J+1)(J-2A)}{2B(2B+1)2C(2C+1)}\right]^{1/2},$$

$$\begin{Bmatrix} A & B & C \\ 1 & C-1 & B-1 \end{Bmatrix} = (-1)^J \left[\frac{J(J+1)(J-2A-1)(J-2A)}{(2B-1)2B(2B+1)(2C-1)2C(2C+1)}\right]^{1/2},$$

$$\begin{Bmatrix} A & B & C \\ 1 & C-1 & B \end{Bmatrix} = (-1)^J \left[\frac{2(J+1)(J-2A)(J-2B)(J-2C+1)}{2B(2B+1)(2B+2)(2C-1)2C(2C+1)}\right]^{1/2},$$

$$\begin{Bmatrix} A & B & C \\ 1 & C-1 & B+1 \end{Bmatrix} = (-1)^J \left[\frac{(J-2B-1)(J-2B)(J-2C+1)(J-2C+2)}{(2B+1)(2B+2)(2B+3)(2C-1)2C(2C+1)}\right]^{1/2},$$

$$\begin{Bmatrix} A & B & C \\ 1 & C & B \end{Bmatrix} = (-1)^J \frac{2[A(A+1)-B(B+1)-C(C+1)]}{[2B(2B+1)(2B+2)2C(2C+1)(2C+2)]^{1/2}}.$$

The following symmetry properties are useful for rearranging the arguments:

$$\begin{Bmatrix} A & B & C \\ D & E & F \end{Bmatrix} = \begin{Bmatrix} D & E & C \\ A & B & F \end{Bmatrix} = \begin{Bmatrix} A & C & B \\ D & F & E \end{Bmatrix} = \begin{Bmatrix} C & A & B \\ F & D & E \end{Bmatrix}$$

etc.

Appendix

2

Four-Dimensional Spherical Harmonics

The following list sets out explicit values of some of the simpler four-dimensional spherical harmonics, as determined from Eq. (2.19):

$$Y_{100} = (2\pi^2)^{-1/2},$$

$$Y_{200} = (2\pi^2)^{-1/2} 2 \cos \chi,$$

$$Y_{300} = (2\pi^2)^{-1/2}(4 \cos^2 \chi - 1),$$

$$Y_{400} = (2\pi^2)^{-1/2}(8 \cos^3 \chi - 4 \cos \chi),$$

$$Y_{210} = -i(2\pi^2)^{-1/2} 2 \sin \chi \cos \theta,$$

$$Y_{310} = -i(2\pi^2)^{-1/2}(6)^{1/2} \sin 2\chi \cos \theta,$$

$$Y_{410} = -i(2\pi^2)^{-1/2} 4(5)^{-1/2} \sin \chi (6 \cos^2 \chi - 1) \cos \theta,$$

$$Y_{320} = -(2\pi^2)^{-1/2}(2)^{1/2} \sin^2 \chi (3 \cos^2 \theta - 1),$$

$$Y_{420} = -(2\pi^2)^{-1/2} 4 \sin^2 \chi \cos \chi (3 \cos^2 \theta - 1),$$

$$Y_{430} = i(2\pi^2)^{-1/2}4(5)^{-1/2} \sin^3 \chi (5 \cos^3 \theta - 3 \cos \theta),$$

$$Y_{211} = i(2\pi^2)^{-1/2}(2)^{1/2} \sin \chi \sin \theta \, e^{i\phi},$$

$$Y_{311} = i(2\pi^2)^{-1/2}2(3)^{1/2} \sin \chi \cos \chi \sin \theta \, e^{i\phi},$$

$$Y_{411} = i(2\pi^2)^{-1/2}(8/5)^{1/2} \sin \chi (6 \cos^2 \chi - 1) \sin \theta \, e^{i\phi},$$

$$Y_{321} = (2\pi^2)^{-1/2}2(3)^{1/2} \sin^2 \chi \cos \theta \sin \theta \, e^{i\phi},$$

$$Y_{421} = (2\pi^2)^{-1/2}4(6)^{1/2} \sin^2 \chi \cos \chi \cos \theta \sin \theta \, e^{i\phi}.$$

References

1. A. R. Edmonds, "Angular Momentum in Quantum Mechanics." Princeton Univ. Press, Princeton, New Jersey, 1960.
2. A. Jucys and A. Bandzaitis, "Teoriya Momenta Kolichestva Dvizheniya v Kvantovoi Mekhanike." Mintis, Vilnius, Lithuania, 1965.
3. D. M. Brink and G. R. Satchler, "Angular Momentum." Oxford Univ. Press, London and New York, 1968.
4. E. Feenberg and G. E. Pake, "Notes on the Quantum Theory of Angular Momentum." Addison-Wesley, Reading, Massachusetts, 1953.
5. A. Messiah, "Mécanique Quantique," Vol. I, Sec. XII-8. Dunod, Paris, 1965.
6. P. A. M. Dirac, "The Principles of Quantum Mechanics." Oxford Univ. Press, New York, 1947.
7. G. Racah, *Phys. Rev.* **62,** 438 (1942).
8. E. P. Wigner, On the matrices which reduce the Kronecker products of representations of S. R. groups. *In* "Quantum Theory of Angular Momentum" (L. C. Biedenharn and H. van Dam, eds.). Academic Press, New York, 1965.
9. M. Rotenberg, R. Bivins, N. Metropolis, and J. K. Wooten, Jr., "The 3-j and 6-j Symbols." Technology Press, MIT, Cambridge, Massachusetts, 1959.
10. E. P. Wigner, "Group Theory." Academic Press, New York, 1959.
11. L. O'Raifeartaigh, Broken symmetry. *In* "Group Theory and its Applications" (E. M. Loebl, ed.). Academic Press, New York, 1968.

12. B. G. Wybourne, "Spectroscopic Properties of Rare Earths." Wiley (Interscience), New York, 1965.
13. K. M. Howell, Tables of 9j-symbols. Res. Rep. 59-2. Univ. of Southampton, Southampton, England, 1959; A. Jucys, Tables of 9j-coefficients for integral values of the parameters with one parameter equal to unity. Comput. Center AS USSR, Moscow, 1968.
14. J. Schwinger, On angular momentum. *In* "Quantum Theory of Angular Momentum" (L. C. Biedenharn and H. van Dam, eds.). Academic Press, New York, 1965.
15. M. E. Rose, "Elementary Theory of Angular Momentum." Wiley, New York, 1957.
16. B. R. Judd, "Operator Techniques in Atomic Spectroscopy." McGraw-Hill, New York, 1963.
17. D. Zwanziger, *Phys. Rev.* **176,** 1480 (1968).
18. A. O. Barut and G. L. Bornzin, *J. Math. Phys.* (*N.Y.*) **12,** 841 (1971).
19. C. J. Joly, "Manual of Quaternions." Macmillan, New York, 1905.
20. H. B. G. Casimir, "Rotation of a Rigid Body in Quantum Mechanics." J. B. Wolters' Uitgevers-Maatschappij N. V., Groningen, den Haag, Batavia, 1931.
21. H. Bateman, *Proc. London Math. Soc.* [2], **3,** 111 (1905).
22. H. R. Hicks and P. Winternitz, *Phys. Rev. D* **4,** 2339 (1971).
23. M. J. M. Hill, *Trans. Cambridge Phil. Soc.* **13,** 273 (1883).
24. J. D. Louck, *J. Mol. Spectrosc.* **4,** 298 (1960).
25. A. P. Stone, *Proc. Cambridge Phil. Soc.* **52,** 424 (1956).
26. L. C. Biedenharn, *J. Math. Phys.* (*N.Y.*) **2,** 433 (1961).
27. A. O. Barut, P. Budini, and C. Fronsdal, *Proc. Roy. Soc. Ser. A* **291,** 106 (1966); C. Fronsdal, *Phys. Rev.* **156,** 1665 (1967); A. O. Barut and H. Kleinert, *Ibid.* **156,** 1541; **157,** 1180; **160,** 1149 (1967).
28. L. Armstrong, Jr., *J. Phys.* (*Paris*) **31,** Suppl. C4, 17 (1970); *Phys. Rev. A* **3,** 1546 (1971).
29. J. D. Talman, "Special Functions." Benjamin, New York, 1968.
30. J. S. Alper, *J. Chem. Phys.* **55,** 3770 (1971); K. G. Kay and J. S. Alper, *Ibid.* **56,** 4243 (1972).
31. C. van Winter, *Physica* (*Utrecht*) **20,** 274 (1954).
32. L. D. Landau and E. M. Lifshitz, "Mechanics," Eqs. 35.1. Addison-Wesley, Reading, Massachusetts, 1960.
33. J. H. Van Vleck, *Rev. Mod. Phys.* **23,** 213 (1951).
34. K. F. Freed, *J. Chem. Phys.* **45,** 4214 (1966).
35. A. Carrington, D. H. Levy, and T. A. Miller, *Advan. Chem. Phys.* **18,** 149 (1970).
36. V. Fock, *Z. Phys.* **98,** 145 (1935).
37. T. Shibuya and C. E. Wulfman, *Proc. Roy. Soc. Ser. A* **286,** 376 (1965). See also J. R. Jasperse, *Phys. Rev. A* **2,** 2232 (1970).
38. P. M. Morse and H. Feshbach, "Methods of Theoretical Physics," Vols. I and II. McGraw-Hill, New York, 1953.
39. B. Podolsky and L. Pauling, *Phys. Rev.* **34,** 109 (1929).
40. W. Pauli, *Z. Phys.* **36,** 336 (1926).
41. W. Pauli, *Ann. Phys.* (Leipzig) **68,** 177 (1922).
42. R. A. Buckingham, Exactly soluble bound state problems. *In* "Quantum Theory" (D. R. Bates, ed.). Academic Press, New York, 1961.
43. E. Teller and H. L. Sahlin, The hydrogen molecular ion and the general theory of electron structure. *In* "Physical Chemistry: An Advanced Treatise" (H. Eyring, ed.), Vol. 5. Academic Press, New York, 1970.
44. D. R. Bates, K. Ledsham, and A. L. Stewart, *Phil. Trans. Roy. Soc. London Ser. A* **246,** 215 (1953).

45. D. R. Bates and R. H. G. Reid, *Advan. At. Mol. Phys.* **4,** 13 (1968).
46. T. E. Sharp, *At. Data* **2,** 119 (1971).
47. M. M. Madsen and J. M. Peek, *At. Data* **2,** 171 (1971).
48. L. Y. Wilson and G. A. Gallup, *J. Chem. Phys.* **45,** 586 (1966).
49. G. Hunter and H. O. Pritchard, *J. Chem. Phys.* **46,** 2146, 2153 (1967).
50. P. I. Pavlik and S. M. Blinder, *J. Chem. Phys.* **46,** 2749 (1967).
51. C. Flammer, "Spheroidal Wave Functions." Stanford Univ. Press, Stanford, California, 1957.
52. G. Jaffé, *Z. Phys.* **87,** 535 (1934).
53. H. A. Erikson and E. L. Hill, *Phys. Rev.* **75,** 29 (1949).
54. C. A. Coulson and A. Joseph, *Int. J. Quantum Chem.* **1,** 337 (1967).
55. C. A. Coulson, Private communication, 1971.
56. W. R. Smythe, "Static and Dynamic Electricity." McGraw-Hill, New York, 1950.
57. C. A. Coulson and P. D. Robinson, *Proc. Phys. Soc. London* **71,** 815 (1958).
58. P. D. Robinson, *Proc. Phys. Soc. London* **71,** 828 (1958).
59. C. A. Coulson and A. Joseph, *Proc. Phys. Soc. London* **90,** 887 (1967).
60. J. W. B. Hughes, *Proc. Phys. Soc. London* **91,** 810 (1967).
61. M. J. Englefield, "Group Theory and the Coulomb Problem." Wiley (Interscience), New York, 1972.
62. Yu. N. Demkov, *Zh. Eksp. Teor. Fiz. Pis'ma Red.* **7,** No. 3, 101 (1968); *JETP Lett.* **7,** 76 (1968).
63. C. E. Wulfman, Dynamical groups in atomic and molecular physics. *In* "Group Theory and Its Applications" (E. M. Loebl, ed.), Vol. II. Academic Press, New York, 1971.
64. E. C. G. Sudarshan, N. Mukunda, and L. O'Raifeartaigh, *Phys. Lett.* **19,** 322 (1965).
65. C. E. Wulfman and Y. Takahata, *J. Chem. Phys.* **47,** 488 (1967).
66. S. Pasternack and R. M. Sternheimer, *J. Math. Phys.* (*N.Y.*) **3,** 1280 (1962).
67. A. Erdelyi, ed., "Higher Transcendental Functions," Vols. I–III. McGraw-Hill, New York, 1953.
67a. B. I. Dunlap, Private communication, 1972.
68. A. O. Barut and W. Rasmussen, *Proc. Phys. Soc. London* (*At. Mol. Phys.*) **6,** 1695 (1973).
69. R. M. Pitzer, C. W. Kern, and W. N. Lipscomb, *J. Chem. Phys.* **37,** 267 (1962).
70. K. G. Kay, H. D. Todd, and H. J. Silverstone, *J. Chem. Phys.* **51,** 2359 (1969).
71. R. J. Buehler and J. O. Hirschfelder, *Phys. Rev.* **83,** 628 (1951).
72. G. M. Copland and D. J. Newman, *Proc. Phys. Soc. London* (*Solid State Phys.*) **5,** 3253 (1972).
73. B. C. Carlson and G. S. Rushbrooke, *Proc. Cambridge Phil. Soc.* **46,** 626 (1950).
74. F. E. Neumann, *J. Reine Angew. Math.* **37,** 21 (1848).
75. K. Ruedenberg, *J. Chem. Phys.* **19,** 1459 (1951).
76. E. W. Hobson, "The Theory of Spherical and Ellipsoidal Harmonics." Cambridge Univ. Press, London and New York, 1931.
77. J. C. Slater, "Quantum Theory of Molecules and Solids," Vol. 1, App. 6. McGraw-Hill, New York, 1963.
78. Y. Sugiura, *Z. Phys.* **45,** 484 (1927).
79. R. L. Matcha, R. H. Pritchard, and C. W. Kern, *J. Math. Phys.* (*N.Y.*) **12,** 1155 (1971).
K. Ruedenberg, *Theor. Chim. Acta* **7,** 359 (1967).
81. G. N. Watson, "A Treatise on the Theory of Bessel Functions." Cambridge Univ. Press, London and New York, 1944.

82. M. J. Lighthill, "Introduction to Fourier Analysis and Generalised Functions." Cambridge Univ. Press, London and New York, 1958.
83. K. H. Hellwege, E. Orlich, and G. Schaack, *Phys. Kondens. Mater.* **4,** 196 (1965).
84. A. J. Freeman and R. E. Watson, *Phys. Rev.* **127,** 2058 (1962).
85. R. J. Birgeneau, M. T. Hutchings, and R. N. Rogers, *Phys. Rev.* **175,** 1116 (1968).
86. J. M. Baker, *Rep. Progr. Phys.* **34,** 109 (1971).
86a. R. A. Sack, *J. Math. Phys. (N.Y.)* **5,** 245 (1964).
87. J. D. Jackson, "Classical Electrodynamics." Wiley, New York, 1962.
88. R. T. Pack and J. O. Hirschfelder, *J. Chem. Phys.* **49,** 4009 (1968).
89. M. Born and J. R. Oppenheimer, *Ann. Phys.* (Leipzig) **84,** 457 (1927).
90. I. Kovacs, "Rotational Structure in the Spectra of Diatomic Molecules." Amer. Elsevier, New York, 1969.
91. J. T. Hougen, *J. Chem. Phys.* **36,** 519 (1962).
92. M. Born and K. Huang, "Dynamical Theory of Crystal Lattices." Oxford Univ Press, London and New York, 1954.
93. L. D. Landau and E. M. Lifshitz, "Quantum Mechanics," p. 123. Pergamon, Oxford, 1965.
94. P. M. Morse, *Phys. Rev.* **34,** 57 (1929).
95. J. L. Dunham, *Phys. Rev.* **41,** 721 (1932).
96. I. Sandeman, *Proc. Roy. Soc. Edinburgh* **60,** 210 (1940).
97. F. Hund, *Z. Phys.* **36,** 657 (1926); **42,** 93 (1927).
98. F. A. Jenkins, *J. Opt. Soc. Am.* **43,** 425 (1953).
99. R. de L. Kronig, "Band Spectra and Molecular Structure." Cambridge Univ. Press, London and New York, 1930.
100. A. O. Barut, "Dynamical Groups and Generalized Symmetries in Quantum Theory." Univ. of Canterbury Press, Christchurch, New Zealand, 1972.
101. V. Heine, "Group Theory in Quantum Mechanics," p. 168. Pergamon, Oxford, 1960.
102. E. Wigner and E. E. Witmer, *Z. Phys.* **51,** 859 (1928).
103. W. Lichten, *Phys. Rev.* **120,** 848 (1960).
104. W. Lichten, *Phys. Rev.* **126,** 1020 (1962).
105. P. R. Brooks, W. Lichten, and R. Reno, *Phys. Rev. A* **4,** 2217 (1971).
106. E. W. Foster and O. Richardson, *Proc. Roy. Soc. Ser. A* **189,** 175 (1947).
107. H. M. Crosswhite, "The Hydrogen Molecule Wavelength Tables of Gerhard Heinrich Dieke." Wiley (Interscience), New York, 1972.
108. P. R. Fontana, *Phys. Rev.* **125,** 220 (1962).
109. L. Y. C. Chiu, *J. Chem. Phys.* **40,** 2276 (1964); *Phys. Rev.* **137,** A384 (1965).
110. L. Y. C. Chiu, *Phys. Rev.* **145,** 1 (1966); **159,** 190 (1967).
111. A. N. Jette and P. Cahill, *Phys. Rev.* **160,** 35 (1967).
112. A. N. Jette, *Phys. Rev. A* **5,** 2009 (1972).
113. H. A. Bethe and E. E. Salpeter, "Quantum Mechanics of One- and Two-Electron Atoms." Academic Press, New York, 1957.
113a. A. N. Jette, *Chem. Phys. Lett.* **25,** 590 (1974).
113b. T. A. Miller and R. S. Freund, *J. Chem. Phys.* **58,** 2345 (1973). See also M. Lombardi, *J. Chem. Phys.* **60,** 4094 (1974).
114. J. K. L. MacDonald, *Proc. Roy. Soc. Ser. A* **136,** 528 (1932).
115. E. Fermi, *Z. Phys.* **60,** 320 (1930).
116. R. A. Frosch and H. M. Foley, *Phys. Rev.* **88,** 1337 (1952).
117. T. M. Dunn, Nuclear hyperfine structure in the electronic spectra of diatomic molecules. *In* "Molecular Spectroscopy: Modern Research" (K. N. Rao and C. W. Mathews, eds.). Academic Press, New York, 1972.

118. K. B. Jefferts, *Phys. Rev. Lett.* **23,** 1476 (1969).
119. N. F. Ramsey, "Molecular Beams." Oxford Univ. Press, London and New York, 1969.
120. S. A. Moszkowski, Models of nuclear structure. *In* "Handbuch der Physik" (S. Flügge, ed.), Vol. 39. Springer-Verlag, Berlin and New York, 1957.
121. B. G. Wicke, R. W. Field, and W. Klemperer, *J. Chem. Phys.* **56,** 5758 (1972).
122. H. E. Radford, *Phys. Rev.* **126,** 1035 (1962).
123. A. Abragam and J. H. Van Vleck, *Phys. Rev.* **92,** 1448 (1953).
124. J. H. Epstein and S. T. Epstein, *Amer. J. Phys.* **30,** 266 (1962); J. C. Nash, *Proc. Phys. Soc. London (At. Mol. Phys.)* **6,** 393 (1973).
125. W. Heitler, "The Quantum Theory of Radiation." Oxford Univ. Press, London and New York, 1954.
126. F. H. Crawford, *Rev. Mod. Phys.* **6,** 90 (1934).
127. G. Herzberg, "Molecular Spectra and Molecular Structure," Vol. I, "Spectra of Diatomic Molecules." Van Nostrand-Reinhold, Princeton, New Jersey, 1950.
128. J. B. Tatum, *Astrophys. J. Suppl. Ser.* **14,** 21 (1967).
129. J. H. Burkhalter, R. S. Anderson, W. V. Smith, and W. Gordy, *Phys. Rev.* **79,** 651 (1950).
130. M. Tinkham and M. W. P. Strandberg, *Phys. Rev.* **97,** 937, 951 (1955).
131. G. Herzberg, *Nature (London)* **163,** 170 (1949).
132. G. Ponzano and T. Regge, Semiclassical limit of Racah coefficients. *In* "Spectroscopic and Group Theoretical Methods in Physics" (F. Bloch, S. G. Cohen, A. de-Shalit, S. Sambursky, and I. Talmi, eds.). Wiley (Interscience), New York, 1968.
133. T. W. Hänsch, M. D. Levenson, and A. L. Schawlow, *Phys. Rev. Lett.* **26,** 946 (1971); M. D. Levenson and A. L. Schawlow, *Phys. Rev. A* **6,** 10 (1972).
134. E. L. Hill and J. H. Van Vleck, *Phys. Rev.* **32,** 250 (1928).
135. F. A. Jenkins, Y. K. Roots, and R. S. Mulliken, *Phys. Rev.* **39,** 16 (1932).
136. R. H. Gillette and E. H. Eyster, *Phys. Rev.* **56,** 1113 (1939).
137. G. H. Dieke and H. M. Crosswhite, *J. Quant. Spectrosc. Radiat. Transfer* **2,** 97 (1962).
138. E. Hulthén, *Z. Phys.* **50,** 319 (1928); R. Rydberg, *Ibid.* **73,** 74 (1931).
139. M. H. Hebb, *Phys. Rev.* **49,** 610 (1936).
140. N. F. Ramsey, *Phys. Rev.* **58,** 226 (1940).
141. W. E. Lamb, *Phys. Rev.* **60,** 817 (1941).
142. D. B. Cook, A. M. Davies, and W. T. Raynes, *Mol. Phys.* **21,** 113, 123 (1971); K. B. MacAdam and N. F. Ramsey, *Phys. Rev. A* **6,** 898 (1972).
143. H. A. Kramers, *Z. Phys.* **53,** 422 (1929).
144. R. H. Pritchard, C. F. Bender, and C. W. Kern, *Chem. Phys. Lett.* **5,** 529 (1970).
145. J. A. Hall, *J. Chem. Phys.* **58,** 410 (1973). See also R. H. Pritchard, C. W. Kern, O. Zamini-Khamiri, and H. F. Hameka, *J. Chem. Phys.* **56,** 5744 (1972); **58,** 411 (1973).
146. J. T. Hougen, The calculation of rotational energy levels and rotational line intensities in diatomic molecules. *Nat. Bur. Stand. (U.S.) Monogr. No.* **115** (1970).
147. R. P. Wayne, *Advan. Photochem.* **7,** 311 (1969).
148. J. H. Van Vleck, *Astrophys. J.* **80,** 161 (1934).
149. J. H. Van Vleck, *Phys. Rev.* **33,** 467 (1929).
150. J. H. Van Vleck, "The Theory of Electric and Magnetic Susceptibilities." Oxford Univ. Press, London and New York, 1932.
151. S. L. Miller and C. H. Townes, *Phys. Rev.* **90,** 537 (1953).
152. C. H. Townes and A. L. Schawlow, "Microwave Spectroscopy." McGraw-Hill, New York, 1955.

Author Index

Numbers in parentheses are reference numbers and indicate that an author's work is referred to although his name is not cited in the text. Numbers in italics show the page on which the complete reference is listed.

Subject Index

K

L

M

N

O

P

Q

R

S

A
B 5
C 6
D 7
E 8
F 9
G 0
H 1
I 2
J 3